Deutsche
Forschungsgemeinschaft

**Entwicklung
der Gentherapie**

**Development
of Gene Therapy**

Deutsche
Forschungsgemeinschaft

Entwicklung der Gentherapie
Development of Gene Therapy

Stellungnahme der Senatskommission
für Grundsatzfragen der Genforschung

Memorandum by the
Senate Commission on Genetic Research

Mitteilung 5/Report 5

WILEY-VCH Verlag GmbH & Co. KGaA

Deutsche Forschungsgemeinschaft
Geschäftsstelle: Kennedyallee 40, 53175 Bonn
Postanschrift: 53170 Bonn
Tel.: +49 228 885-1
Fax: +49 228 885-2777
postmaster@dfg.de
www.dfg.de

Das vorliegende Werk wurde sorgfältig erarbeitet. Dennoch übernehmen Autoren, Herausgeber und Verlag für die Richtigkeit von Angaben, Hinweisen und Ratschlägen sowie für eventuelle Druckfehler keine Haftung.

Bibliografische Information der Deutschen Nationalbibliothek
Die Deutsche Nationalbibliothek verzeichnet diese Publikation in der Deutschen Nationalbibliografie; detaillierte bibliografische Daten sind im Internet über <http://dnb.d-nb.de> abrufbar.

ISBN: 978-3-527-31907-7

© 2007 WILEY-VCH Verlag GmbH & Co. KGaA, Weinheim

Gedruckt auf säurefreiem und chlorfrei gebleichtem Papier.
Alle Rechte, insbesondere die der Übersetzung in andere Sprachen, vorbehalten. Kein Teil dieses Buches darf ohne schriftliche Genehmigung des Verlages in irgendeiner Form – durch Photokopie, Mikroverfilmung oder irgendein anderes Verfahren – reproduziert oder in eine von Maschinen, insbesondere von Datenverarbeitungsmaschinen, verwendbare Sprache übertragen oder übersetzt werden. Die Wiedergabe von Warenbezeichnungen, Handelsnamen oder sonstigen Kennzeichen in diesem Buch berechtigt nicht zu der Annahme, dass diese von jedermann frei benutzt werden dürfen. Vielmehr kann es sich auch dann um eingetragene Warenzeichen oder sonstige gesetzlich geschützte Kennzeichen handeln, wenn sie nicht eigens als solche markiert sind.

All rights reserved (including those of translation into other languages). No part of this book may be reproduced in any form – by photoprinting, microfilm, or any other means – nor transmitted or translated into a machine language without written permission from the publishers. Registered names, trademarks, etc. used in this book, even when not specifically marked as such, are not to be considered unprotected by law.

Umschlaggestaltung und Typographie: Dieter Hüsken
Wiley Bicentennial Logo: Richard J. Pacifico
Satz: Hagedorn Kommunikation, Viernheim
Druck: Strauss GmbH, Mörlenbach
Buchbinder: Litges & Dopf GmbH, Heppenheim

Printed in the Federal Republic of Germany

Inhalt

	Vorwort	IX
1	Zusammenfassung	1
2	Einführung und kurzer historischer Abriss	3
3	Klinische Anwendung: Erfolge und Rückschläge	5
4	Aktuelle Situation und weiterer Forschungsbedarf	13
5	Rechtliche und ethische Aspekte	19
6	Schlussfolgerungen und Empfehlungen	23
7	Weiterführende Literatur	25
8	Glossar	27
9	Mitglieder der Arbeitsgruppe „Gentherapie", die die vorliegende Stellungnahme verfasst haben	33
10	Mitglieder der Senatskommission für Grundsatzfragen der Genforschung	35

Contents

	Preface	41
1	Summary	43
2	Introduction and brief historic outline	45
3	Clinical application: successes and setbacks	47
4	Present situation and further research needs	53
5	Legal and ethical considerations	59
6	Conclusions and recommendations	63
7	References	65
8	Glossary	67
9	Members of the Working Group on "Gene Therapy", who prepared the present statement	73
10	Members of the Senate Commission on Genetic Research	75

Vorwort

Nach etwas mehr als zehn Jahren legt die Senatskommission für Grundsatzfragen der Genforschung eine zweite Stellungnahme zur Gentherapie vor. Wie kaum eine andere neuartige Therapieform hat die Gentherapie seit dem Ende der 1980er Jahre immer wieder zu kontrovers geführten Diskussionen bezüglich ihres therapeutischen Potenzials und den damit verbundenen gesundheitlichen Risiken und ethischen Problemen geführt. Mittlerweile konnten jedoch umfangreiche experimentelle Arbeiten und erste klinische Anwendungen die bis dahin nur zu vermutenden therapeutischen Möglichkeiten sowie deren Nebenwirkungen erheblich konkretisieren. Wie die nun vorliegende Stellungnahme zeigt, ist die Umsetzung der frühen Heilsversprechungen deutlich langsamer vorangekommen als erhofft, doch auch überzogene Risikoeinschätzungen haben sich glücklicherweise nicht bewahrheitet. Therapieerfolge wie bei den angeborenen Immunschwächekrankheiten zeigen, dass die somatische Gentherapie eine brauchbare Therapieoption sein kann. Die hierbei beobachteten und zum Teil fatalen Nebenwirkungen zeigen aber auch die derzeit bestehenden Risiken der Therapie. In dieser Problematik unterscheidet sich die somatische Gentherapie jedoch nicht grundsätzlich von anderen therapeutischen Ansätzen. Es gilt: Vor und während der klinischen Erprobung und der Anwendung müssen Nutzen und Risiken abgewogen werden, immer auch im Vergleich mit den jeweiligen alternativ existierenden Therapiemöglichkeiten und auf der Basis solider Erkenntnisse aus experimentellen und klinischen Beobachtungen. Hierfür sind neben einem stets aktuellen Wissensstand der enge Kontakt und offene Austausch mit

Vorwort

Grundlagenwissenschaftlern und klinisch tätigen Forschern und Forscherinnen, aber auch mit den beratenden Ethikkommissionen und Zulassungsbehörden essenziell. Alle Wissenschaftler sind aufgefordert, diesen offenen Diskurs zu üben, ebenso wie die Öffentlichkeit aufgefordert ist, sich durchaus kritisch, aber unvoreingenommen an dieser Diskussion zu beteiligen. Diese Stellungnahme richtet sich daher sowohl an die Wissenschaft wie auch an die interessierte Öffentlichkeit, um den derzeitigen Stand und die weiteren Perspektiven der somatischen Gentherapie aufzuzeigen.

Da sich die somatische Gentherapie mittels retroviraler Vektoren trotz einer sich abzeichnenden Konsolidierung weiterhin in einem stark experimentellen Stadium befindet, warnt die Senatskommission zu Recht vor einer breiten Anwendung zum jetzigen Zeitpunkt. Neben der weiteren Therapieoptimierung besteht nach wie vor ein erheblicher Forschungsbedarf bezüglich unseres Verständnisses der beobachteten und zu erwartenden Nebenwirkungen. Um so erfreulicher ist es, dass die Deutsche Forschungsgemeinschaft (DFG) aktuell mit der Förderung eines speziellen Schwerpunktprogramms zur Untersuchung des Zelleintritts und der Persistenz von gentherapeutischen Vektoren auch einen wesentlichen Beitrag zum Verständnis der Mechanismen leistet, die zu möglichen Nebenwirkungen bei der somatischen Gentherapie führen können. Auch von den ausländischen Gutachtern zu diesem Schwerpunkt wurde uns bestätigt, dass wir in Deutschland eine Reihe international führender Gruppen und vielversprechenden Nachwuchs auf diesem Gebiet besitzen. Daher ist es umso wichtiger, diese Wissenschaftlerinnen und Wissenschaftler neben der finanziellen Förderung auch durch die Bereitstellung geeigneter klinischer und wissenschaftlicher Strukturen, aber auch durch das Aufzeigen sichtbarer beruflicher Perspektiven zu unterstützen.

Vorwort

Ich möchte an dieser Stelle der Senatskommission für Grundsatzfragen der Genforschung und den Autoren dieser Stellungnahme dafür danken, dass sie die Entwicklung der somatischen Gentherapie umfassend dargelegt haben und die notwendige Diskussion auf der Basis des aktuellen Erkenntnisstandes und der zukünftigen Perspektiven begleiten.

Bonn, Dezember 2006

Professor Dr. Ernst-Ludwig Winnacker
Präsident der Deutschen Forschungsgemeinschaft

1 Zusammenfassung

Seit die Deutsche Forschungsgemeinschaft (DFG) im Jahr 1995 eine erste Stellungnahme zum Thema Gentherapie veröffentlichte, hat sich dieses Forschungsfeld enorm weiterentwickelt. Erste klinische Erfolge haben das therapeutische Potenzial, aber auch die Risiken dieser Therapieoption aufgezeigt. Nicht anders als andere therapeutische Ansätze auch erfordert die klinische Anwendung gentherapeutischer Arzneimittel eine sorgfältige Abschätzung von Nutzen und Risiken. Nach wie vor befindet sich dieses Forschungsfeld in einem Pilotstadium und ist, wie wenige medizinische Forschungsfelder sonst, zwingend auf die enge Zusammenarbeit und den ständigen Austausch von Grundlagenwissenschaftlern und Klinikern verschiedener Disziplinen angewiesen. Dies erfordert einen iterativen Optimierungsprozess, wobei, anders als in der klassischen klinischen Forschung, neue Ergebnisse aus dem Labor direkt in die klinische Prüfung transportiert und dabei gewonnene Erkenntnisse wiederum zur Weiterentwicklung im Labor genutzt werden.

Zur erfolgreichen Weiterentwicklung der Gentherapie besteht mittelfristig Forschungsbedarf insbesondere im Bereich der Entwicklung effizienter und sicherer Vektoren sowie bezüglich der molekularen Untersuchung der Wirkungen und Nebenwirkungen der Gentherapie im Tiermodell und im Rahmen experimenteller klinischer Studien. Dabei muss auch das Risikoprofil potenziell einzusetzender therapeutischer Gene stärker berücksichtigt werden. Die Anwendung der Gentherapie ist dem therapeutischen und präventiven Bereich vorbehalten. Für das Gendoping oder den kosmetischen Bereich wird keine vertretbare

1 Zusammenfassung

Anwendung gesehen. Die gegenwärtigen rechtlichen Rahmenbedingungen sind ausreichend; die derzeitige Erfassung von Gentransfer-Studien in einem zentralen Register soll fortgeführt werden. Insgesamt haben die letzten zehn Jahre mit ersten Erfolgen bei Immunschwächekrankheiten das therapeutische Potenzial der Gentherapie gezeigt und vielversprechende Therapieansätze auch für eine Reihe von monogenetischen Erbkrankheiten und für die erworbenen Krankheiten Krebs und HIV-Infektion/AIDS geliefert.

2 Einführung und kurzer historischer Abriss

Die **Gentherapie** ist definiert als das Einbringen von Genen in Gewebe oder Zellen mit dem Ziel, durch die Expression und Funktion dieser Gene therapeutischen oder präventiven Nutzen zu erlangen. Der Vorgang des Einbringens von Genen in Zellen wird als **Gentransfer** bezeichnet. Hierfür benötigt man ein Vehikel, welches das Gen trägt, den **Vektor**. Bei der Gentherapie handelt es sich also um eine medizinische Behandlung mit Gentransfer-Arzneimitteln, die im Arzneimittelgesetz definiert sind. Der erwünschte Gentransfer bezieht sich ausschließlich auf somatische Zellen (somatische Gentherapie). Der Keimbahngentransfer ist in Deutschland gesetzlich verboten.

In den letzten beiden Jahrzehnten erhielten nur wenige medizinische Forschungsgebiete so viel Aufmerksamkeit wie die somatische Gentherapie. Die Entdeckungen der Molekularbiologie und Genetik, die im Jahr 2001 in der Entschlüsselung weiter Teile des menschlichen Genoms einen Höhepunkt fanden, schufen die Voraussetzungen für die Therapie mit Genen. In der frühen Phase der Gentherapie Anfang der 1990er Jahre wurde dieser Sachverhalt teilweise unkritisch und zu euphorisch diskutiert. Dies führte zu einer unrealistischen Erwartungshaltung. Zwar wurden die Methoden des Gentransfers in den letzten Jahren stark verbessert, und erste klinisch erfolgreiche Gentherapien wurden bei Patienten durchgeführt. Dennoch muss auch heute betont werden, dass die Entwicklung ausgereifter Gentherapie-Verfahren für viele ansonsten nicht behandelbare Krankheiten viele Jahre dauern wird, wenn auch bei einzelnen Gentherapie-Ansätzen der Erfolg mittelfristig absehbar erscheint.

2 Einführung und kurzer historischer Abriss

Die ersten gut dokumentierten Gentherapie-Studien wurden Anfang der 1990er Jahre begonnen. Bis 2005 wurden schätzungsweise mehr als 1100 Gentherapie-Studien weltweit durchgeführt, davon ein Drittel in Europa mit einem Schwerpunkt in Deutschland. In China wurden im Jahr 2003 sowie im November 2005 erste Gentherapie-Arzneimittel für die Behandlung bestimmter maligner Tumoren zugelassen. Ein erster europäischer Zulassungsantrag für ein Gentherapie-Arzneimittel zur Behandlung eines aggressiven Hirntumors wurde im Jahr 2005 bei der Europäischen Arzneimittelagentur (EMEA) eingereicht.

Trotz weiterhin großer Schwierigkeiten in der technischen Umsetzung können heute die Erfolge der Gentherapie zweifelsfrei belegt werden. Beispielsweise gelangen in den letzten fünf Jahren bei Patienten mit schweren angeborenen Immunschwächekrankheiten erfolgreiche, kausale Therapien, von denen die lebensbedrohlich erkrankten Patienten sichtbar profitierten (siehe Abschnitt 3). Bei drei der mit retroviralen Vektoren behandelten Kinder einer Studie trat allerdings drei Jahre nach der Behandlung ein leukämieähnliches Krankheitsbild auf, das zunächst ausschließlich auf eine Nebenwirkung der verwendeten Vektoren in Kombination mit der Grunderkrankung zurückgeführt wurde, möglicherweise aber zusätzlich mit der Überexpression des verwendeten Zielgens zusammenhängt. Dies sowie der Tod eines Patienten in den USA im Jahr 1999 durch eine sehr hohe, systemisch verabreichte Dosis adenoviraler Vektoren waren tragische Ereignisse, die in der Öffentlichkeit als Rückschlag für die Gentherapie gesehen wurden. Gleichwohl gelten für die Gentherapie dieselben Grundsätze wie für andere medizinische Eingriffe: Wirksame Verfahren sind mit potenziellen Nebenwirkungen verbunden. Nebenwirkungen können durch Verbesserung der Verfahren reduziert werden, sobald die zugrunde liegenden Mechanismen verstanden sind. Für jede Indikation und jedes Verfahren muss das Verhältnis von Wirkung zu Nebenwirkungen (therapeutischer Index) in sorgfältigen präklinischen und klinischen Studien bestimmt werden. Deutsche Wissenschaftler haben wichtige Beiträge zu diesem Gebiet von der Grundlagenforschung der Vektor-Wirt-Interaktion bis hin zu klinischen Studien geleistet und unter anderem im Jahr 2006 über die Korrektur einer schweren Immunschwäche bei erwachsenen Patienten mittels Gentherapie berichtet.

3 Klinische Anwendung: Erfolge und Rückschläge

Die Gentherapie bietet über bisherige Ansätze hinaus einen neuen Weg der Therapie mit hohem Innovationspotenzial, da hier Gene als Arzneimittel verwendet werden, während die konventionelle Arzneimittelentwicklung chemische Stoffe, Produkte von Mikro-Organismen oder Proteine verwendet. In der Praxis zeigte sich allerdings rasch, dass die anfänglichen Prognosen zur Gentherapie den hohen Entwicklungsaufwand deutlich unterschätzt hatten.

Die Wahl geeigneter Vektoren ist für die Wirksamkeit einer Gentherapie entscheidend. Dabei können unterschiedliche Therapie- oder Präventionsziele verschiedene Vektoren erfordern. Die Auswahl hängt zum Beispiel davon ab, ob der Gentransfer im Patienten (*in vivo*) oder in der Zellkulturschale (*ex vivo* beziehungsweise *in vitro*) stattfindet, da diese Verfahren unterschiedliche Anforderungen an die Sicherheit und Zielgenauigkeit des Vektors stellen. Die Entwicklung von verbesserten Vektoren für die Gentherapie ist weiterhin eine der zentralen Aufgaben für die Forschung.

Gentransfer-Vektoren für die somatische Gentherapie müssen im Wesentlichen folgende Eigenschaften haben:

- Sie müssen bestimmte Zellen des Menschen effizient modifizieren können.
- Sie müssen eine ausreichend starke und ausreichend langfristige Genexpression gewährleisten können.
- Sie müssen ein möglichst geringes Risikoprofil im Hinblick auf den gewünschten Behandlungsansatz aufweisen.

3 Klinische Anwendung: Erfolge und Rückschläge

In klinischen Gentherapie-Studien werden derzeit vielfach nicht vermehrungsfähige virale Vektoren eingesetzt, die von Retroviren, Adenoviren, Adeno-assoziierten Viren und Pockenviren abgeleitet sind. Hierzu wurden aus dem Virusgenom Abschnitte entfernt oder inaktiviert, die zur Replikation notwendig sind, und durch das therapeutische Gen ersetzt. Nach Einbringen der Vektoren in Helferzelllinien, die für die Virusbildung notwendige Funktionen bereitstellen, entstehen defekte Viren, die zum Gentransfer geeignet sind, sich aber außerhalb von Helferzellen nicht mehr vermehren können. Daneben kommt Plasmid-DNA in reiner Form oder mit weiteren Reagenzien gemischt als nichtviraler Vektor zum Einsatz. In der Krebstherapie und Impfung finden bedingt vermehrungsfähige Viren mit therapeutischen oder präventiven Genen ihren klinischen Einsatz.

Die Erkrankungsgruppen, die bisher hauptsächlich in klinischen Studien zur Gentherapie untersucht wurden, sind Krebserkrankungen, monogene Erbkrankheiten, Infektionskrankheiten (insbesondere HIV/AIDS) und kardiovaskuläre Krankheiten. Dabei stellen Krebserkrankungen mit über 60 Prozent den größten Anteil. Die meisten klinischen Gentherapie-Studien befinden sich in sehr frühen klinischen Phasen, nur wenige haben die klinische Prüfung der Phase III erreicht oder den Nachweis einer klinischen Wirksamkeit erbracht. Es muss davon ausgegangen werden, dass viele der bisher durchgeführten oder derzeit laufenden klinischen Gentherapie-Studien der Phasen I und II noch nicht zu einem routinemäßig einsetzbaren Arzneimittel führen werden, da es sich hierbei um Pilotansätze zur Behandlung sehr seltener Erkrankungen handelt.

Ein Nachweis der klinischen Wirksamkeit einer Gentherapie konnte insbesondere in Studien zur Behandlung schwerer Immundefekte erbracht werden. Dies sind Studien zu den angeborenen kombinierten Immundefekten („Severe Combined Immunodeficiency", X-SCID; Arbeitsgruppe Cavazzano-Calvi und Fischer, Paris, sowie Arbeitsgruppe Thrasher, London), zum Adenosin-Deaminase-Mangel (ADA-SCID; Arbeitsgruppe Auiti und Bordignon, Mailand) und zur chronischen Granulomatose (CGD; Arbeitsgruppe Grez und Hölzer, Frankfurt/Main). Erste Hinweise auf klinische Wirksamkeit der Gentherapie ergaben sich weiterhin bei der Verwendung des CD40-Liganden bei chronischer lymphatischer Leukämie (Arbeitsgruppe Kipps, San Diego), beim Transfer

3 Klinische Anwendung: Erfolge und Rückschläge

des GM-CSF-Gens beim malignen Melanom (Arbeitsgruppe Lattime, Philadelphia) sowie bei der Therapie der Hämophilie B (Faktor IX; Arbeitsgruppe McKay und High, Stanford/Philadelphia).

Etwa zehn Jahre intensiver Forschungs- und Entwicklungsarbeit vieler Gruppen waren erforderlich, um eine ausreichend große Zahl von Zellen außerhalb des menschlichen Körpers so mit Genvektoren zu behandeln, dass nach Rückgabe dieser genetisch modifizierten Zellen im Patienten ein therapeutischer Erfolg erzielt werden konnte. Erwartungsgemäß wurden diese ersten Erfolge bei der Therapie monogenetischer Erbkrankheiten mittels retroviral modifizierter Blutstamm- beziehungsweise Vorläuferzellen erzielt und hier insbesondere bei den Immundefekt-Syndromen X-SCID und ADA-SCID. Dabei kommen vom murinen Leukämievirus abgeleitete, nicht vermehrungsfähige Vektoren *ex vivo* zum Einsatz, die zur weitgehend zufälligen Integration des therapeutischen Gens in ein Chromosom der jeweiligen Wirtszelle führen. Die genannten Erkrankungen bieten besonders günstige Voraussetzungen für die Gentherapie, da ein therapeutischer Effekt bereits bei Modifikation einer begrenzten Zahl von Zielzellen erreicht wird, die genetisch modifizierten Zellen im Organismus einen Wachstumsvorteil haben oder dieser durch Vorbehandlung der Patienten erreicht wird und die mit dem Gentransfer-Vektor modifizierten Zellen aufgrund der Immunschwäche nicht abgestoßen werden.

Etwa drei Jahre nach der ersten erfolgreichen Behandlung von zehn Patienten mit X-SCID in der Studie des Pariser Necker-Hospitals haben drei dieser Patienten akute T-Zell-Leukämien als Nebenwirkung der Gentherapie entwickelt und einer der Patienten ist an dieser Erkrankung verstorben. In einer modellhaften internationalen Kooperation unter Beteiligung deutscher Wissenschaftler wurden wesentliche Fortschritte bei der Aufklärung der molekularen Ursachen dieser Nebenwirkung erzielt. Es stellte sich heraus, dass die verwendeten retroviralen Genvektoren durch den Einbau in das Genom der behandelten T-Zellen zelluläre Proto-Onkogene aktiviert hatten und so zur Auslösung dieser Krebserkrankungen beigetragen haben. Zusätzlich ergaben neuere Untersuchungen Zweifel an der Unbedenklichkeit des verwendeten Korrekturgens, da bei Langzeitstudien im Tiermodell Lymphome auftraten, die nicht unbedingt mit dem verwendeten Vektor verknüpft zu sein scheinen. Entsprechende Nebenwirkungen sind in einer ähnlichen Studie an X-SCID-Patienten des Londoner Great Ormond Street-Kinderkrankenhauses bisher nicht aufgetreten, sodass insgesamt die Erfolgsbilanz dieser Gentherapie-Studien positiv ist. Allerdings ist der in der Pariser Studie erkannte prinzipielle Nachteil der verwendeten Vektoren hinsichtlich des Langzeitverlaufs aus Mangel an klinischen Erfahrungswerten derzeit nicht sicher einschätzbar. Auch bei der im Jahr 2006 berichteten Behandlung von erwachsenen Patienten mit CGD mittels retroviraler Vektoren in Frankfurt/Main fand sich ein häufiger Einbau des Vektorgenoms in zellzyklusaktivierende Gene, wobei dies in diesem Fall vermutlich zum therapeutischen Erfolg beigetragen hat (durch Vermehrung der erfolgreich behandelten Zellen im Organismus). Zwischenzeitlich ist allerdings einer der in dieser Studie behandelten Patienten an einer Infektionskomplikation nach weitgehendem Funktionsverlust der therapierten Zellen gestorben, während die anderen weiterhin von der therapeutischen Wirkung profitieren. Zusammengenommen zeigen diese Ergebnisse deutlich, dass noch erheblicher Forschungsbedarf bezüglich der Zusammenhänge von therapeutischer Effizienz und Nebenwirkungen bei der Gentherapie besteht.

Die auf dem Gebiet der Gentherapie tätigen Forscher haben von Beginn an die öffentliche Diskussion gesucht und auch bei sehr frühen klinischen Prüfungen einen vollständigen Einblick in die Therapierisiken und -nebenwirkungen gegeben. Dabei ist in der Öffentlichkeit nicht ausreichend wahrgenommen worden, dass die Gentherapie tödlich verlaufender Krankheiten, wie beispielsweise angeborener monogener Immunschwächekrankheiten, trotz der oben beschriebenen Leukämien keine höhere Nebenwirkungsrate als vergleichbare konventionelle Therapieformen aufweist. Bisher trat bei drei der insgesamt 28 behandelten Patienten mit X-SCID eine Leukämie als Nebenwirkung der verwendeten Vektoren auf, woran einer der Patienten verstarb. Dies entspricht einer Nebenwirkungsrate von etwa zehn Prozent bei einer Mortalitätsrate von vier Prozent. Bei der konventionellen Therapie derselben Krankheiten durch Knochenmark- beziehungsweise Blutstammzelltransplantation von HLA-identischen Familienspendern liegt die Mortalitätsrate bei zehn Prozent oder höher. Bei der für die Gentherapie bisher gewählten Gruppe von Kindern ohne HLA-identischen Spender außerhalb der Familie ist

3 Klinische Anwendung: Erfolge und Rückschläge

die Mortalitätsrate mit etwa 30 Prozent sogar noch deutlich höher. Allerdings muss darauf hingewiesen werden, dass die genannten Zahlen auf einer schmalen Datenbasis beruhen. Die Leukämie als Nebenwirkung war zwar bereits vor Einsatz der Gentherapie als mögliches Risiko bekannt, allerdings war die Wahrscheinlichkeit ihres Eintretens unklar und wurde als gering eingeschätzt. Nach gewissenhafter Prüfung und Abwägung des Risikos haben sich alle Beteiligten zur Therapie entschieden, weil ohne Gentherapie eine länger dauernde Korrektur der Zellfunktionen nicht möglich war und das Risiko der Grunderkrankung dasjenige des therapeutischen Eingriffs deutlich überstieg.

Angesichts neuer Erkenntnisse über die Ursachen der Leukämie wurden in der Zwischenzeit neue, sicherere Vektoren entwickelt, die das Risiko der Aktivierung zellulärer Onkogene deutlich vermindern sollen. Außerdem wurden sensitive präklinische Modelle und diagnostische Methoden zur Toxizitätsbestimmung beschrieben. So können die molekularen Mechanismen schwerer Nebenwirkungen in Zukunft oft vermieden oder früher erkannt werden. Neueste Daten zeigen darüber hinaus, dass die toxische Wirkung des neu eingeführten Gens im Fall X-SCID möglicherweise unterschätzt worden ist, sodass neben dem Vektor zukünftig auch das therapeutische Gen in die Beurteilung des Risikoprofils stärker einbezogen werden muss. Aufgrund neuerer Untersuchungen ist weiterhin deutlich geworden, dass die bisher verwendeten Vektoren zwar nur sehr selten zu schweren Nebenwirkungen führen, aber relativ häufig Auswirkungen auf die Expression zellulärer Gene haben. Diese Vektoren beeinflussen also Wachstum und Funktion genmodifizierter Zellen auch im ansonsten gesunden Organismus. Dies zeigte sich auch in der oben beschriebenen CGD-Studie. Das Risiko schwerer Nebenwirkungen ist wahrscheinlich von vielen weiteren Faktoren und von der Grunderkrankung abhängig; dies kann im experimentellen Modell zwar analysiert werden, wird aber letztlich erst in der klinischen Erprobung eindeutig erkennbar. Sowohl bei der Vektorentwicklung als auch bei der klinischen Umsetzung besteht also weiterhin erheblicher Forschungsbedarf.

Die genannten Erfolge der Gentherapie wurden durch eine stark verbesserte Effizienz des Gentransfers ermöglicht. So können zum Beispiel blutbildende Zellen heute mit hoher Effizienz (>50 Prozent) genetisch modifiziert werden. Das Erreichen effektiver therapeutischer Wirkspiegel ging jedoch einher mit einer höheren Wahrscheinlichkeit symptomatischer Nebenwirkungen. Dies bedeutet gleichzeitig, dass – wie bei anderen Arzneimitteltherapien – bei weiterer Dosiserhöhung der bisher verwendeten Vektoren Nebenwirkungen wahrscheinlicher und ausgeprägter werden dürften. Dosisfindung und Toxizitätsermittlung gehört seit jeher untrennbar zur Entwicklung pharmazeutischer Wirksubstanzen, und auch hier macht die Gentherapie keine Ausnahme. Durch genaue Erforschung der molekularen Ursachen sind vielversprechende Ansätze erkennbar geworden, die das Nutzen-Risiko-Profil der nächsten Generationen von Genvektoren deutlich verbessern werden. Für eine Reihe von monogenetischen Erbkrankheiten und die erworbenen Erkrankungen Krebs und HIV-Infektion/AIDS bietet die Gentherapie daher mehr denn je innovative und verfolgenswerte Therapieansätze. Ziel der HIV-Gentherapie ist dabei insbesondere das Einbringen schützender Gene (die zum Beispiel den Viruseintritt verhindern) in Blutstammzellen des Patienten, wobei die erfolgreich behandelten Zellen dann möglichst langfristig HIV-resistente Immunzellen produzieren und so das Immunsystem funktionell erhalten sollen.

Im Gegensatz zur oben dargestellten Methode bei unbehandelt tödlich verlaufenden Krankheiten erfordert die Anwendung von Gentherapie-Vektoren zur Expression von Antigenen als Impfstoff den Einsatz von Vektoren und Verfahren mit sehr geringem Nebenwirkungsrisiko. Auch in diesem Fall muss sich die Nutzen-Risiko-Analyse auf die gegenwärtig eingesetzten Impfstoffe gegen Infektionskrankheiten beziehen, deren Nebenwirkungsrisiko im Promillebereich oder darunter liegt. Für diese Anwendung kommen daher beispielsweise nicht-virale Vektoren oder vermehrungsunfähige virale Vektoren zum Einsatz, die nur eine vorübergehende und lokale Zellmodifizierung, aber keine chromosomale Integration bewirken. Das Risiko dieser Vektoren wird als sehr gering eingeschätzt, allerdings ist auch die Gentransfer-Effizienz niedriger und die Genexpression nur vorübergehend. In jedem Fall ist vor der klinischen Prüfung von Gentransfer-Arzneimitteln, wie bei anderen Arzneimitteln auch, eine auf den einzelnen Ansatz bezogene verantwortungsvolle Nutzen-Risiko-Analyse vorzunehmen.

Aufgrund der Risiken gentherapeutischer Verfahren mit inserierenden Vektorsystemen sowie der für ihre Entwicklung not-

3 Klinische Anwendung: Erfolge und Rückschläge

wendigen Kostenaufwendungen ist eine Anwendung retroviral modifizierter Zellen am Menschen auf absehbare Zeit nur zur Behandlung schwerwiegender Erkrankungen unter genauer Abwägung des jeweiligen Nutzen-Risiko-Verhältnisses zulässig. Vektorsysteme, die keine dauerhaften Veränderungen hervorrufen – und die sich demnach von anderen pharmazeutischen Wirkstoffen nicht prinzipiell unterscheiden – sind dagegen nach umfangreichen Sicherheitstests des jeweiligen Vektors und transgenen Produkts auch bei nicht lebensbedrohlichen Erkrankungen einsetzbar. Selbst mit nachgewiesenermaßen sicheren Vektoren ist jedoch eine nicht-medizinisch indizierte leistungsverbessernde Anwendung der Gentherapie, etwa im Leistungssport (Gendoping), aus ethischen und medizinischen Gründen nicht vertretbar.

4 Aktuelle Situation und weiterer Forschungsbedarf

Die Forschung zur Entwicklung einer erfolgreichen Gentherapie ist beispielhaft für andere Bereiche der *translationalen Forschung*, in der Erkenntnisse aus der biomedizinischen Grundlagenforschung direkt in die klinische Anwendung übertragen werden sollen. Dieser Bereich der Medizin steht besonderen Herausforderungen und Schwierigkeiten gegenüber, und sein Erfolg hängt in hohem Maße von einem funktionierenden Dialog zwischen Grundlagenwissenschaft (zum Beispiel zur Vektorentwicklung und Optimierung) und angewandter klinischer Forschung ab. Von besonderer Bedeutung für die erfolgreiche Entwicklung der Gentherapie-Forschung in Deutschland ist das Zusammenwirken von Grundlagenforschern und in der Behandlung der Zielkrankheit erfahrenen Ärzten in einem Team. Wie aus den oben dargestellten Ergebnissen bisheriger klinischer Studien deutlich wird, erfordert die Weiterentwicklung Erfolg versprechender Gentherapie-Ansätze einen iterativen Optimierungsprozess, der, anders als in der bisherigen klinischen Forschung, neue Erkenntnisse aus dem Labor in die Klinik transportiert, dort testet, eine Hypothese erhärtet oder verwirft und anschließend erneute präklinische Optimierungen nach sich zieht. Dies hat selbstverständlich nach strengen ethischen Kriterien zu erfolgen.

In Deutschland ist besonders durch die gezielte Förderung des Bundesministeriums für Bildung und Forschung (BMBF) und die Projektförderung durch die DFG in den letzten zehn Jahren eine substanzielle Zahl erfolgreicher und interdisziplinär arbeitender Gentherapie-Gruppen etabliert worden. Vor allem ist es gelungen, eine Reihe jüngerer Naturwissenschaftler und Mediziner

4 Aktuelle Situation und weiterer Forschungsbedarf

aus den USA nach Deutschland zurückzuholen. Insbesondere auf den Gebieten der Vektorentwicklung und der Genominsertionsanalyse retroviraler Vektoren, aber auch bei der Durchführung klinischer Gentherapie-Studien, haben deutsche Wissenschaftler eine international anerkannte Position errungen. Die Pionierzeit der raschen Erfolge in der Gentherapie-Forschung geht zu Ende. Mehr und mehr ist eine systematische Grundlagenforschung zur Lösung der erkannten Probleme notwendig. Bei entsprechender Förderung sollte die deutsche Forschungstradition der systematischen und detaillierten Analyse eine gute Grundlage bieten, langfristig die erreichte Position zu erhalten und auszubauen. Forschungsbedarf besteht dabei insbesondere in den Bereichen (i) Verbesserung von Effizienz und Sicherheit der Gentransfer-Vektoren, (ii) Optimierung der Spezifität der verwendeten Viren für definierte Zielzellen zum Einsatz in der *in vivo*-Gentherapie, (iii) Untersuchung des Verbleibs genmodifizierter Zellen im Patienten und (iv) Erforschung der molekularen Ursachen von Nebenwirkungen. Neben diesen unmittelbaren Forschungsthemen spielt für die Entwicklung der Gentherapie auch der Aspekt der Vektorproduktion und -tests eine wichtige Rolle, der bei unzureichender Unterstützung limitierend für den weiteren Fortschritt werden könnte.

Bei den ersten klinischen Erfolgen der Gentherapie bei der Behandlung von X-SCID bestanden besondere Voraussetzungen: Die von diesen Patienten entnommenen defekten Knochenmarkzellen erhalten durch den therapeutischen Gentransfer einen Selektionsvorteil, da sie nach Rücktransplantation in den Patienten auf die natürlichen Signale des Körpers ansprechen und sich vermehren können. Durch diese „geheilten" Zellen wird das Immunsystem regeneriert, und dieser Selektionsvorteil erlaubt auch bei Behandlung einer relativ geringen Zahl von Zellen außerhalb des Organismus (*ex vivo*) einen therapeutischen Erfolg. Analog trat bei der gentherapeutischen Behandlung der CGD anscheinend ein Selektionsvorteil für die *ex vivo* therapierten Zellen durch Insertion des Vektorgenoms in für das Zellwachstum bedeutsame Gene auf. Bei den meisten Erkrankungen erwartet man jedoch keinen Selektionsvorteil für die behandelten Zellen. Darüber hinaus ist – wie oben gesagt – eine weitere Erhöhung der Vektordosis möglicherweise mit erhöhten Komplikationen verbunden. Demzufolge besteht weiterhin Forschungsbedarf bei der Entwicklung effizienter und gleichzeitig sicherer Vektoren sowie bei der Untersuchung des Risikoprofils therapeutisch einzusetzender Gene, zunächst *in vitro* und im Tiermodell und – bei entsprechendem Erfolg – im Rahmen klinischer Studien. Durch ein besseres Verständnis der Faktoren, die zur Vervielfältigung der verwendeten Viren beitragen, können Verbesserungen erzielt werden, wobei mittelfristig eine weitere Optimierung durch Verknüpfung viraler Funktionskomplexe aus verschiedenen Viren sowie in Kombination mit nicht-viralen Systemen erwartet werden kann. So könnte es möglich werden, erwünschte Eigenschaften unterschiedlicher Vektorsysteme miteinander zu verknüpfen und möglicherweise unerwünschte Eigenschaften auszuschalten. Vor dem Hintergrund des sich entwickelnden Gebiets der synthetischen Biologie scheint die synthetische Herstellung eines Gentransfer-Vektors realisierbar.

Neben der Tatsache, dass nur relativ wenige Zellen bei der *ex vivo*-Gentherapie behandelt werden können, besteht der weitere Nachteil darin, dass – mit Ausnahme der Zellen des Blutes – die meisten Körperzellen nicht einfach für die Therapie in Zellkultur entnommen werden können. Daher wird eine Verabreichung des Gentransfer-Vektors im Patienten (*in vivo*) angestrebt. Dabei werden die verabreichten Vektoren in ihrer Konzentration im Blut sehr rasch verdünnt und kommen mit vielen anderen, nicht von der jeweiligen Erkrankung betroffenen Zellarten in Berührung. Um therapeutisch wirksam zu werden und möglichst geringe Nebenwirkungen in anderen Zellen auszulösen, müssen die *in vivo* applizierten Vektoren daher viel spezifischer und viel effektiver beim Eindringen in ihre jeweiligen Zielzellen und in ihrer therapeutischen Wirkung in diesen Zellen sein als beim *ex vivo*-Gentransfer. Die Erhöhung der Spezifität der Vektoren für therapeutisch bedeutsame Zielzellen des Organismus und der Effektivität ihres Eindringens und ihrer Wirkung in diesen Zellen stellt daher ein weiteres zentrales Forschungsthema und eine unabdingbare Voraussetzung für eine breitere klinische Anwendung der Gentherapie dar.

Forschungsbedarf besteht weiterhin bei der Untersuchung der molekularen Ursachen von Nebenwirkungen der Gentherapie. Gerade dieses Thema ist durch das Auftreten von drei Leukämiefällen bei 28 erfolgreich behandelten Patienten mit X-SCID hoch aktuell geworden. Das Risiko von Nebenwirkungen, die durch den weitgehend ungerichteten Einbau des retroviralen Vektorgenoms in das Genom der Wirtszelle verursacht werden kann, war ursprünglich als sehr gering eingeschätzt worden. Durch den zufälligen Einbau kann die Funktion von Genen, die das normale Zellwachstum steuern, gestört werden, was zu ungerichtetem Wachstum und Tumoren führen kann. Trotz erheblicher Fortschritte sind die Ursachen des unerwartet gehäuften Auftretens dieser Nebenwirkung bei den X-SCID-Patienten noch nicht völlig geklärt. Die Aufklärung der Ursachen von Nebenwirkungen wird nicht nur zum Verständnis der zugrunde liegenden molekularen Mechanismen, sondern auch zu effektiveren und sichereren Gentransfer-Protokollen und Vektoren für die klinische Anwendung führen. Neuere Untersuchungen zeigen, dass eine Veränderung der Zellproliferation durch Vektorinsertion möglicherweise häufiger auftritt, diese aber nicht notwendigerweise zu Tumoren führt. Die Erforschung der zugrunde liegenden Ursachen und Bedingungen ist für die weitere Entwicklung dieses Ansatzes der Gentherapie von entscheidender Bedeutung. Neben dem ungerichteten Einbau des Vektors kann auch das verwendete therapeutische Gen selbst bei der Tumorentstehung eine Rolle spielen. Erste Studien diesbezüglich zeigten, dass diese Effekte erst mit einiger Verzögerung auftreten und sich daher in Kurzzeitexperimenten möglicherweise nicht manifestieren. Es gilt daher, langfristige präklinische Beobachtungen durchzuführen und Tiermodelle zu entwickeln, die entsprechende Nebenwirkungen möglichst schnell anzeigen.

Ein zentraler Faktor für die klinische Anwendung der Gentherapie ist die Herstellung der erforderlichen Menge des Gentransfer-Arzneimittels entsprechend den regulatorischen Vorgaben („Good Manufacturing Practice" bei Herstellung des Prüfarzneimittels (GMP), „Good Laboratory Practice" bei pharmakologisch-toxikologischen Tests (GLP), „Good Clinical Practice" bei der klinischen Prüfung (GCP)). Die Entwicklung der Gentherapie liegt derzeit weitgehend in den Händen von akademischen Forschungsgruppen und kleinen Biotechnologie-Unternehmen. Es steht außer Frage, dass universitäre Gruppen in der Regel nicht in der Lage sind, eine Vektorherstellung entsprechend GMP-Vorgaben zu erreichen. In anderen Ländern sind unterschiedliche Wege hinsichtlich der GMP-Produktion von Gentherapie-Vektoren oder genetisch modifizierten Zellen begangen worden. In den USA haben sich universitäre Einrichtungen, Biotechnologiefirmen und das NIH engagiert. Die französische Stiftung „l'Association Française contre les Myopathies" hat von 1997 bis 2003 ein „Gene Vector Production Network" etabliert, um die Nutzung und Modifizierung von Gentherapie-Vektoren für Forschungszwecke zu erleichtern. Eine der Ideen war, daraus ein europäisches Netzwerk auch für die GMP-Produktion aufzubauen, was aber nicht verwirklicht wurde. In Großbritannien hat das Department of Health für den Zeitraum 2003 bis 2008 vier Millionen GBP zur Produktion von Gentherapie-Vektoren für Gentherapie-Studien innerhalb des NHS (National Health Systems) bereitgestellt. In Deutschland bestehen derzeit Möglichkeiten zur GMP-gerechten Herstellung von Vektoren für die Gentherapie bei Biotechnologie-Unternehmen, einigen Pharmafirmen sowie am Helmholtz-Zentrum für Infektionsforschung in Braunschweig. Grundsätzlich können Vektorproduktion und -tests sowohl über Etablierung gemeinsamer Produktionsstätten im öffentlichen Bereich als auch durch hierauf spezialisierte mittelständische Unternehmen (häufig als Ausgründungen universitärer Gruppen) als Serviceanbieter realisiert werden, wobei in jedem Fall eine ausreichende Finanzierung hierfür erforderlich ist. Diese Kosten können derzeit nur selten durch die Projektförderung der interdisziplinär arbeitenden akademischen Forschungsgruppen von Ärzten und Naturwissenschaftlern getragen werden. Unter Berücksichtigung der Tatsache, dass Gentherapie nur im Wechselspiel zwischen Forschung und Entwicklung auf der einen und klinischer Prüfung auf der anderen Seite erfolgreich fortentwickelt werden kann, stellt der Aspekt von Vektorproduktion und -tests einen wichtigen Standortfaktor für dieses Forschungsgebiet dar.

Zusammenfassend ist festzustellen, dass grundlegende Fortschritte in der erfolgreichen Anwendung der Gentherapie nur auf der Basis intensiver grundlagenorientierter, präklinischer und klinischer Forschung und Prüfung sowie in stetiger Kommunikation dieser Teilbereiche erzielt werden können. Diesem Forschungsbedarf trägt die DFG unter anderem durch die Einrichtung des Schwerpunktprogramms „Mechanismen des Zelleintritts und der Persistenz von Genvektoren" im Jahr 2005 Rechnung. Das Ziel dieses Schwerpunktprogramms ist die fachübergreifende Untersuchung der biologischen Sicherheit des Eintritts und der Persistenz von viralen und nicht-viralen Gentransfer-Vektoren

4 Aktuelle Situation und weiterer Forschungsbedarf

mit wissenschaftlichem Fokus auf den Zellen des blutbildenden und lymphatischen Systems. Dieser stärker grundlagenorientierte DFG-Schwerpunkt ergänzt sich mit anderen nationalen („Innovative Therapien" des BMBF) und internationalen (CLINIGEN der EU) Förderinitiativen, die zur klinischen Anwendung überleiten. Da jede dieser Förderinitiativen jeweils nur einen Teil des für die translationale Forschung notwendigen Spektrums der Grundlagen- und der klinischen Forschung abdeckt, ist eine enge Absprache zwischen den verschiedenen Förderorganisationen unabdingbar, um den disziplinübergreifenden Gruppen von Wissenschaftlern und Klinikern ein Förderangebot in der benötigten Breite zu bieten. Das gemeinsam von BMBF und DFG durchgeführte Programm zur Förderung klinischer Studien zeigt, wie eine erfolgreiche Zusammenarbeit in diesem Bereich aussehen kann. Einen alternativen Weg bietet das Programm der Klinischen Forschergruppen der DFG, in denen auf der Basis einer klar umrissenen thematischen Fokussierung klinisch relevante Forschungsfelder in den Kliniken, auch unter Mitarbeit von Grundlagenwissenschaftlern, schwerpunktmäßig verankert werden. Die enge Interaktion zwischen Grundlagenforschern und Klinikern innerhalb eines solchen Verbunds gewährleistet optimale Bedingungen für Erfordernisse der translationalen Forschung. Als Beispiel hierfür sei die Klinische Forschergruppe „Stammzell-Therapie" an der Medizinischen Hochschule in Hannover genannt, in der auch gentherapeutische Ansätze in die Klinik getragen werden.

5 Rechtliche und ethische Aspekte

Der Keimbahngentransfer ist in Deutschland durch das Embryonenschutzgesetz wegen des mit ihm verbundenen unabsehbaren Risikopotenzials aus guten Gründen verboten. Auf ihn wird daher im Folgenden nicht näher eingegangen.

Die somatische Gentherapie wirft dagegen im Vergleich zu anderen innovativen Therapien keine grundlegend anderen oder neuen ethischen oder rechtlichen Probleme auf. Vor einer Erstanwendung am Menschen sind die mit ihr verbundenen Risiken tierexperimentell abzuklären. Die Kommission Somatische Gentherapie des Wissenschaftlichen Beirats der Bundesärztekammer hat zudem in mehreren Grundsatzentscheidungen von individuellen Heilversuchen mit Gentransfer-Arzneimitteln abgeraten, weil eine Therapieentwicklung rational nur durch Erkenntnisgewinn aufgrund der Anwendung an einer Reihe von Probanden oder Patienten im Rahmen einer klinischen Prüfung möglich erscheint.

Die Herstellung gentechnisch veränderter Organismen im Labor sowie die Errichtung und der Betrieb gentechnischer Anlagen unterliegen der Anmelde- oder Genehmigungspflicht nach den §§ 8ff. des Gentechnikgesetzes (GenTG). Das Gentechnikgesetz erfasst dagegen nicht die Anwendung von gentechnisch veränderten Organismen am Menschen. Auf eine klinische Prüfung ist daher vor allem das Arzneimittelgesetz (AMG), auf die spätere Zulassung die EG Verordnung 726/2004 anzuwenden. Allerdings kann der Behandlungsraum im Rahmen einer klinischen Prüfung eine gentechnische Anlage im Sinne des Gentechnikgesetzes darstellen.

5 Rechtliche und ethische Aspekte

Wie jede andere im Versuchsstadium befindliche Arzneimitteltherapie bedarf die klinische Prüfung von Gentransfer-Arzneimitteln einer freiwilligen und selbst bestimmten Einwilligung der Studienteilnehmer; der Einwilligung muss eine ausreichende Aufklärung vonseiten eines Arztes vorangegangen sein, die insbesondere auch Hinweise auf die Neuartigkeit der Maßnahme und die zu erwartenden beziehungsweise zu befürchtenden Risiken umfassen muss. Vor und bei Durchführung der klinischen Prüfung ist zudem eine Nutzen-Risiko-Abwägung erforderlich, in deren Rahmen die Risiken der Arzneimittelanwendung und das Schutzbedürfnis der Zielgruppe (Patienten oder gesunde Probanden) einerseits gegenüber dem möglichen Nutzen für die Zielgruppe und der Bedeutung des Arzneimittels für die Medizin andererseits abzuwägen sind. Diese Abwägung führt beispielsweise dazu, dass Gentransfer-Arzneimittel mit geringem Risiko als vorbeugende Impfstoffe gegen Infektionskrankheiten an gesunden Probanden angewandt werden, während andere Gentransfer-Arzneimittel zur Therapie letaler Krankheiten wie beispielsweise bestimmter Hirntumoren lediglich an konventionell austherapierten Patienten mit einer Lebenserwartung von nur noch wenigen Monaten erprobt werden.

Mit Inkrafttreten der 12. Novelle des Arzneimittelgesetzes (AMG) im Jahr 2004, durch welche die europäische GCP-Direktive (Richtlinie 2001/20/EG) umgesetzt wurde, ist speziell festgelegt, welche Richtlinien und Vorschriften für die Herstellung und Entwicklung bis hin zur Marktzulassung von Gentherapie- oder Gentransfer-Arzneimitteln (beide Begriffe sind weitgehend synonym) anzuwenden sind und den gesicherten Stand der wissenschaftlichen Erkenntnis definieren. Nach § 4 Abs. 9 des Arzneimittelgesetzes gehören dieser Arzneimittelgruppe zum einen virale und nicht-virale Gentransfer-Vektoren, Plasmid-DNA und onkolytische Viren für den *in vivo*-Gentransfer, zum anderen *ex vivo* genetisch modifizierte Zellen an.

Zulassungspflicht besteht in der Gentherapie zum einen für industriell oder gewerblich hergestellte Individualrezepturen. Betroffen sind beispielsweise von Unternehmen oder auch Blutbanken hergestellte Gentransfer-Arzneimittel, die nach einem einheitlichen Muster unter Verwendung derselben Genfähre und desselben therapeutischen Gens genetisch modifizierte Zellen enthalten und an Ärzte zur Anwendung im Rahmen der Präven-

5 Rechtliche und ethische Aspekte

tion, Therapie oder *in vivo*-Diagnostik bei einem bestimmten Patienten abgegeben werden sollen. Zulassungspflicht besteht zum anderen für vorab hergestellte Gentransfer-Arzneimittel für die *in vivo*-Verabreichung bei vielen Patienten, wie zum Beispiel virale Vektoren. Die Zulassung kann auf der Basis von Ergebnissen klinischer Prüfungen der Phasen I bis III bei der Europäischen Arzneimittelagentur (EMEA) beantragt werden.

Vor Beginn einer klinischen Prüfung sind die positive Stellungnahme der zuständigen Ethikkommission und die Genehmigung des Paul-Ehrlich-Instituts (PEI) notwendig. Ethikkommissionen ziehen, soweit keines ihrer Mitglieder einschlägige Expertise aufweist, bei der Bewertung von Anträgen auf klinische Gentherapie-Prüfung externe Experten zur Beratung hinzu. Die von 1994 bis 2005 übliche Beratung der Ethikkommissionen durch die Kommission Somatische Gentherapie des Wissenschaftlichen Beirats der Bundesärztekammer wurde durch einen Beschluss des Vorstands der Bundesärztekammer bis auf Weiteres ausgesetzt.

Durch die 12. AMG-Novelle wurden die Genehmigungs- beziehungsweise Stellungnahmefristen des Paul-Ehrlich-Instituts beziehungsweise der zuständigen Ethikkommission gesetzlich festgelegt, was die Verfahren beschleunigt. Bei der klinischen Prüfung von Gentransfer-Arzneimitteln, die genetisch veränderte Organismen, zum Beispiel virale Vektoren und bedingt vermehrungsfähige Viren oder Mikro-Organismen beinhalten, umfasst die Genehmigung der klinischen Prüfung durch das Paul-Ehrlich-Institut auch die erforderliche Freisetzungsgenehmigung. Gleiches gilt für die Zulassung gentherapeutischer Arzneimittel durch die EMEA.

Da Gentransfer-Arzneimittel eine neue Klasse von Arzneimitteln darstellen, die einem anhaltenden Entwicklungsprozess unterliegen, können regulatorische Leitfäden in der Regel nur generelle Hinweise geben. Über Studien zum Nachweis von Qualität, Sicherheit und Wirksamkeit eines gegebenen Gentransfer-Arzneimittels muss zumeist im Einzelfall entschieden werden. Dabei sind kleine Biotechnologie-Unternehmen oder akademische Forschungseinrichtungen häufig mit neuen Fragen konfrontiert. Hier kann der Antragsteller beim Paul-Ehrlich-Institut eine Beratung im Vorfeld der Antragstellung auf klinische Prüfung in Anspruch nehmen. Dabei ist die Geheimhaltung der Daten und die Vertraulichkeit des Inhalts der Beratungsgespräche gesetzlich garantiert.

5 Rechtliche und ethische Aspekte

Seit dem Jahr 2004 wurde eine europäische Datenbank (EudraCT) bei der EMEA installiert, welche den zuständigen Behörden, der Europäischen Kommission und der EMEA die notwendigen Informationen über klinische Prüfungen in allen europäischen Mitgliedstaaten gibt. Dieses Register ist allerdings nicht öffentlich zugänglich. In Deutschland existiert zusätzlich das „Deutsche Register für somatische Gentransferstudien" (DeReG). Dieses Register wurde 2001 auf Betreiben der Deutschen Gesellschaft für Gentherapie (DG-GT) und der Kommission Somatische Gentherapie in Freiburg eingerichtet und durch das BMBF gefördert. Es erfasst Informationen, die in keinem anderen derzeit verfügbaren internationalen Studienregister abgefragt werden können. So werden in dem Freiburger Register auch Nebenwirkungen bei einzelnen Patienten aus kleinen Phase-I-Studien registriert. Außerdem kann durch dieses Register die Öffentlichkeit bei Bedarf (auftretende Nebenwirkungen oder Erfolge) schnell und zuverlässig informiert werden. Der Erhalt eines solchen öffentlich zugänglichen Registers erscheint daher sinnvoll, um die Transparenz im Bereich Gentherapie zu erhöhen. Die DFG fordert daher weiterhin von Antragstellern, dass vor der Bewilligung von klinischen Forschungsprojekten die Registrierung der Gentherapie-Studie im DeReG nachgewiesen wird. Ob das geplante Nationale Studienregister die genannten Aufgaben in analoger Weise erfüllen kann, ist derzeit offen und wird sich erst in der weiteren Umsetzung zeigen.

Während man den gentherapiespezifischen rechtlichen Rahmen als ausreichend betrachten kann, gelten die grundsätzlichen regulatorischen und strukturellen Probleme in gleicher Form für die gentherapeutische Forschung im klinischen Umfeld, wie sie allgemein für die akademisch betriebene klinische Forschung in Deutschland herrschen. Die Rahmenbedingungen für die klinische Forschung in Deutschland insgesamt günstiger zu gestalten würde daher auch die Situation der im Bereich der klinischen Umsetzung der Gentherapie forschenden Wissenschaftler erheblich verbessern. Die hier bestehenden Probleme und auch entsprechende Lösungsvorschläge sind in der DFG-Denkschrift zur Klinischen Forschung aus dem Jahr 1999 sowie in den „Zehn Eckpunkten zur klinischen Forschung" aus dem Jahr 2004 beschrieben, die in weiten Teilen noch nichts von ihrer Aktualität eingebüßt haben.

6 Schlussfolgerungen und Empfehlungen

- Seit der ersten Stellungnahme der DFG zur Gentherapie aus dem Jahr 1995 hat die somatische Gentherapie insbesondere bei monogenetisch bedingten Immunschwächekrankheiten eindeutige therapeutische Erfolge gezeigt. Bei anderen potenziellen Anwendungen befindet sie sich noch in eher frühen Stadien.

- Wie jede medizinische Behandlung beinhaltet auch die Gentherapie Risiken, die genau überwacht und hinsichtlich ihrer molekularen Ursachen aufgeklärt werden müssen. Die klinische Anwendung erfordert eine sorgfältige Abschätzung von Nutzen und Risiken für die jeweiligen Indikationen. Dabei muss die öffentliche Diskussion von Erfolgen und Rückschlägen auch die Prognose der Grunderkrankung und alternative Therapieoptionen berücksichtigen.

- Die Anwendung der Gentherapie ist dem therapeutischen und präventiven Bereich vorbehalten. Eine Anwendung für das Gendoping oder den kosmetischen Bereich wird abgelehnt.

- Die Therapie mit retroviralen Vektoren sollte, solange keine sicheren Viren vorliegen, nur bei Krankheiten ohne alternative therapeutische Option zum Einsatz kommen.

- Die Gentherapie hat weiterhin einen hohen Forschungsbedarf. Dabei sollte Grundlagenforschung in direkter, interdisziplinärer Verknüpfung mit Untersuchungen im Tiermodell und mit klinischen Studien durchgeführt werden. Förderprogramme wie die Klinischen Forschergruppen bieten eine Grundlage

6 Schlussfolgerungen und Empfehlungen

hierfür und sollten zunehmend durch Mittel zur Unterstützung translationaler Forschung seitens der Fakultäten ergänzt werden. Darüber hinaus muss die notwendige Finanzierung der aufwändigen Vektorproduktion in geeigneten, nach GMP-Richtlinien genehmigten Anlagen beziehungsweise durch kommerzielle Anbieter gesichert sein.

- Aktueller Forschungsbedarf besteht insbesondere bezüglich der Entwicklung effizienter und sicherer Vektoren für die Anwendung *in vitro* und *in vivo* inklusive erhöhter Spezifität für definierte Zielzellen sowie bezüglich der molekularen Untersuchung der Wirkungen und Nebenwirkungen. Dabei muss auch das Risikoprofil klinisch einzusetzender therapeutischer Gene berücksichtigt werden.

- Die bestehenden rechtlichen Rahmenbedingungen zur Gentherapie sind ausreichend.

- Die Erfassung von Gentransfer-Studien der Phase I und II in einem zentralen Register hat sich bewährt und ist nach wie vor sinnvoll. Die Registrierung sollte auch weiterhin Voraussetzung für eine Förderung durch die DFG sein.

7 Weiterführende Literatur

Baum C, Dullmann J, Li Z, Fehse B, Meyer J, Williams DA, von Kalle C: Side effects of retroviral gene transfer into hematopoietic stem cells. Blood. 2003 Mar 15; 101 (6): 2099–2114. Epub 2003 Jan 2. Review.

Guidance for Industry Gene Therapy Clinical Trials – Observing Participants for Delayed Adverse Events. FDA CBER. www.fda.gov/Cber/gene.htm.

Hallek M, Buening H, Ried MU, Hacker U, Kurzeder C, Wendtner C-M: Grundlagen der Gentherapie. Internist 2001; 42: 1306–1313.

Kay MA, Glorioso JC, Naldini L: Viral vectors for gene therapy: the art of turning infectious agents into vehicles of therapeutics. Nat Med. 2001 Jan; 7 (1): 33–40. Review.

von Kalle C, Baum C, Williams DA: Lenti in red: progress in gene therapy for human hemoglobinopathies. J Clin Invest. 2004 Oct; 114 (7): 889–891.

Nabel GJ: Genetic, cellular and immune approaches to disease therapy: past and future. Nat Med. 2004 Feb; 10 (2): 135–141. Review.

O'Connor TP, Crystal RG: Genetic medicines: treatment strategies for hereditary disorders. Nat Rev Genet. 2006 Apr; 7 (4): 261–176. Review.

Nienhuis AW, Dunbar CE, Sorrentino BP: Genotoxicity of retroviral integration in hematopoietic cells. Mol Ther. 2006 Jun; 13 (6): 1031–1049. Epub 2006 Apr 19.

DFG-Denkschrift „Klinische Forschung". 1999. www.dfg.de/aktuelles_presse/reden_stellungnahmen/archiv/denkschrift_klinische_forschung.html

7 Weiterführende Literatur

Internetadressen

Deutsches Register für somatische Gentransferstudien (DeReG)
   www.dereg.de
Deutsche Gesellschaft für Gentherapie e.V. (DG-GT)
   www99.mh-hannover.de/kliniken/zellth/dggt
European Society for Gene Therapy (ESGT)
   www.esgt.org
Paul-Ehrlich-Institut
   www.pei.de

8 Glossar

AAV-Vektoren: Adeno-assoziierte Viren, die in der Gentherapie eingesetzt werden. Sie sind in der Regel nicht mit humanen Krankheiten assoziiert, bilden stabile Partikel und infizieren auch ruhende Zellen, wo sie sich stabil ins Genom integrieren können. AAV-Partikel haben jedoch nur eine sehr begrenzte Aufnahmekapazität für fremde Gene. Damit sich das AAV vermehren kann, benötigt es ein zweites Virus (so genanntes Helfervirus, meist ein Adeno- oder Parvovirus).

ADA-SCID: Eine angeborene schwere kombinierte Immunerkrankung (SCID, severe combined immunodeficiency), bei der durch einen Gendefekt das Enzym Adenosin-Deaminase (ADA) fehlt. Als Folge kann der Körper ein für die weißen Blutkörperchen giftiges Protein nicht abbauen und die für die Immunabwehr wichtigen T-Lymphozyten reifen im Knochenmark nicht oder nur in zu geringer Zahl heran. Die von dieser Krankheit betroffenen Kinder sind allen Krankheitserregern fast vollkommen schutzlos ausgesetzt und überleben trotz Behandlung und einem Leben unter sterilen Bedingungen nur selten ihre Kindheit.

Adenosin-Deaminase-Mangel: s. ADA-SCID.

Adenovirale Vektoren: Adenoviren sind unter anderem für Erkältungskrankheiten beim Menschen verantwortlich. Replikationsdefekte Vektoren haben eine relativ hohe genetische Aufnahmekapazität, können in höheren Dosierungen jedoch zu starken Immunantworten nach der Verabreichung führen.

8 Glossar

AMG: Gesetz über den Verkehr mit Arzneimitteln (Arzneimittelgesetz). Eine aktuelle Fassung finden Sie unter www.gesetze-im-internet.de/amg_1976/index.html.

BMBF: Bundesministerium für Bildung und Forschung.

CGD: Chronic Granulomatous Disease, s. chronische Granulomatose.

Chromosomale Integration: Fester Einbau viraler oder eingeführter Fremdgene in die Chromosomen des Empfängers.

Chronische Granulomatose: Genetisch bedingte Störung der Sauerstoffradikalbildung von Phagozyten. Durch die gestörte Phagozytenfunktion sind die Patienten stark infektionsanfällig und leiden an entzündlichen Erkrankungen.

DeReG: Deutsches Register für somatische Gentransferstudien (www.dereg.de). Dieses Register ist öffentlich zugänglich.

DG-GT: Deutsche Gesellschaft für Gentherapie e.V.

EMEA: European Medicines Agency, Arzneimittelzulassungsbehörde der EU.

EudraCT: EU-weites Register für Klinische Studien der EMEA. Dieses Register ist nicht öffentlich zugänglich.

Ex vivo-Gentransfer: Gentransfer-Verfahren, bei dem die Zielzellen, in der Regel des blutbildenden Systems, zunächst aus dem Körper isoliert werden, um dann mit dem Vektor genetisch verändert und gegebenenfalls angereichert zu werden. Anschließend werden diese Zellen wieder dem Körper verabreicht.

GCP: s. Good Clinical Practice.

Genexpression: Umsetzung der genetischen Information, meist in Form von Proteinen, zur Bildung von Zellstrukturen und Signalen.

Genfähre: Andere Bezeichnung für einen (Gen-)Vektor.

GenTG: Gentechnikgesetz. Eine aktuelle Fassung finden Sie unter www.gesetze-im-internet.de/gentg/index.html.

Gentherapie: Heilansatz durch Einbringen von Genen in Gewebe oder Zellen mit dem Ziel, durch die Expression und Funktion dieser Gene therapeutischen oder präventiven Nutzen zu erlangen.

8 Glossar

Gentransfer: Der methodische Vorgang des Einbringens von Genen in Zellen.

GLP: s. Good Laboratory Practice.

GM-CSF: Granulozyten-Makrophagen koloniestimulierender Faktor. Ein so genanntes Zytokin, welches das Wachstum von Makrophagen anregt und damit eine Immunreaktion gegen bestimmte Formen von Hautkrebs induzieren kann.

GMP: s. Good Manufacturing Practice.

Good Clinical Practice: Internationale Regeln zur Vorbereitung und Durchführung klinischer Studien nach ethischen und praktischen Aspekten auf der Basis der aktuellen wissenschaftlichen Erkenntnis. Weitere Details (Englisch) finden Sie unter www.emea.eu.int/pdfs/human/ich/013595en.pdf.

Good Laboratory Practice: Internationale Regeln und Standards zur Qualitätssicherung der organisatorischen Prozesse und Bedingungen von nicht-klinischen Gesundheits- und Umweltprüfungen. Weitere Details (Englisch) finden Sie unter http://ec.europa.eu/enterprise/chemicals/legislation/glp/index_en.htm.

Good Manufacturing Practice: Internationale Regeln und Standards zur Qualitätssicherung in der Herstellung von medizinischen Produkten und Wirkstoffen. Weitere Details (Englisch) finden Sie unter www.emea.eu.int/Inspections/GMPhome.html.

In vivo-Gentransfer: Im Gegensatz zum *ex vivo*-Gentransfer (s. o.) werden hier die Genvektoren direkt in den Körper des Patienten eingebracht. In Abhängigkeit von der Zellspezifität des benutzten Vektors erfolgt dann die Infektion beziehungsweise der Einbau des Fremdgens mehr oder weniger zielgerichtet in bestimmte Zelltypen.

Keimbahngentransfer: Gentransfer in Keimzellen (Ei- beziehungsweise Samenzellen oder deren Vorläufer). Veränderungen im Erbgut würden auch auf nachfolgende Generationen vererbt. Der Keimbahntransfer ist in Deutschland gesetzlich verboten.

Klinische Prüfung der Phasen I, II, III und IV: Studien zur Wirksamkeit und Toxizität von Arzneimitteln am Menschen. Diese Prüfungen unterliegen strengen Bestimmungen. In der Phase I wird zunächst an einer kleinen Zahl von gesunden Probanden die Toxizität beziehungsweise Verträglichkeit von neuen Wirkstoffen geprüft. Aufbauend auf den Ergebnissen der Phase I wird in Phase II an einer größeren Zahl von Studienteilnehmern die optimale Dosierung festgestellt. In der Phase III wird die eigentliche Wirkung an einer für eine statistisch valide Auswertung ausreichend großen Zahl von Patienten mit bestimmten Ein- und Ausschlusskriterien bestimmt. Hierzu gehört gegebenenfalls der Vergleich mit einem Scheinmedikament ohne wirksame Inhaltsstoffe (Placebo). Erst auf der Basis einer erfolgreichen Phase-III-Studie ist die Zulassung eines neuen Arzneimittels möglich. Danach können die Wirkungen einer neuen Therapie in ihrer zugelassenen Anwendung weiter untersucht beziehungsweise beobachtet werden. Man spricht dann von einer so genannten Phase-IV-Studie.

Monogene Erbkrankheiten: Krankheiten, die durch die Veränderung eines einzelnen Gens hervorgerufen werden.

Onkogen: Gen, das üblicherweise eine Rolle in der Zellzyklusregulation spielt und dessen Aktivierung durch Mutation zur Krebsentwicklung beiträgt oder diese auslöst.

Onkolytische Viren: Viren, die gezielt Tumorzellen infizieren und ausschalten können.

Plasmid-DNA: DNA, die nicht in ein Genom eingebaut ist, sondern als eigenständige, ringförmige Struktur in einer Zelle vorliegt. Diese wird bei der Zellteilung in der Regel nicht verdoppelt und verliert sich so nach mehreren Zellteilungen – es sei denn, die Plasmid-DNA wird dauerhaft in das Genom eingebaut.

Proto-Onkogen: Gen, das durch eine Mutation zu einem Onkogen (s. o.) verändert werden kann.

Retrovirale Vektoren: Genvektoren, die sich von Retroviren ableiten. Retroviren gehören zu den RNA-Viren, ihr RNA-Genom wird allerdings in DNA umgeschrieben und dauerhaft in das Genom einer Zelle eingebaut. Retrovirale Vektoren auf Basis der murinen Leukämieviren infizieren viele verschiedene Zelltypen zum Teil mit sehr hoher Effizienz. Allerdings können sie nicht teilungsaktive Zellen (zum Beispiel Nervenzellen) nicht infizieren. Dies kann jedoch durch Verwendung lentiviraler Vektoren auf Basis des HIV erreicht werden.

8 Glossar

Somatische Gentherapie: Anwendung des Gentransfers auf somatische Zellen (s. u.). Genetische Veränderungen werden hierbei nicht an die Nachkommen weitergegeben.

Somatische Zellen: Körperzellen, deren genetische Information nicht an nachfolgende Generationen weitervererbt werden kann. Sie bilden den Großteil der menschlichen Zellen, lediglich Keimzellen (Ei- und Samenzellen) können Erbinformationen auf die nächste Generation übertragen und bilden die so genannte Keimbahn (s. o.).

Therapeutischer Index: Der therapeutische Index (auch therapeutische Breite oder therapeutischer Quotient) eines Arzneimittels beschreibt das Verhältnis seiner therapeutischen zu seiner toxischen Dosis. Je größer der therapeutische Index ist, umso ungefährlicher ist ein Arzneimittel.

T-Zell-Leukämie: Blutkrebs, bei dem die aus der Kontrolle geratene Regulation der Zellvermehrung weißer Blutzellen (T-Zellen) zu einer Überschwemmung von Blut und Lymphsystem mit entarteten Zellen führt.

Vektor: Ein Vehikel, das ein therapeutisches Gen in die Zellen des Patienten trägt. Neben verschiedenen, meist vermehrungsunfähigen Viren kommt Plasmid-DNA (s. o.) in reiner Form oder mit weiteren Reagenzien gemischt als nicht-viraler Vektor zum Einsatz.

X-SCID: Eine angeborene schwere kombinierte Immunerkrankung (SCID, severe combined immunodeficiency). Durch eine Mutation eines Gens für einen gemeinsamen Baustein mehrerer verschiedener Typen von Interleukinrezeptoren können keine Abwehrzellen des Immunsystems gebildet werden, sodass betroffene Patienten, meist Kinder, hochanfällig für Infektionen sind. Das zugrunde liegende Gen ist auf dem X-Chromosom lokalisiert, daher die Bezeichnung X-SCID.

9 Mitglieder der Arbeitsgruppe „Gentherapie", die die vorliegende Stellungnahme verfasst haben

Als Mitglieder der Senatskommission für
Grundsatzfragen der Genforschung:

Prof. Dr. Hans-Georg Kräusslich – Vorsitzender –	Ruprecht-Karls-Universität Heidelberg Abteilung Virologie Im Neuenheimer Feld 324 69120 Heidelberg
Prof. Dr. Claus R. Bartram	Ruprecht-Karls-Universität Heidelberg Institut für Humangenetik Im Neuenheimer Feld 366 69120 Heidelberg
Prof. Dr. Jochen Taupitz	Universität Mannheim Institut für Deutsches, Europäisches und Internationales Medizinrecht Schloss 68131 Mannheim

9 Mitglieder der Arbeitsgruppe „Gentherapie"

Als externe Experten:

Prof. Dr. Klaus Cichutek

Paul-Ehrlich-Institut (PEI)
Paul-Ehrlich-Straße 51–59
63225 Langen

Prof. Dr. Charles Coutelle

Imperial College London
Faculty of Life Sciences
Division of Cell and Molecular Biology
Wolfson Biochemistry Building
Exhibition Road, South Kensington
London SW7 2AY
Großbritannien

Prof. Dr. Michael Hallek

Universität zu Köln
Klinik I für Innere Medizin
Hämatologie und Onkologie
Joseph-Stelzmann-Straße 9
50931 Köln

Prof. Dr. Christof von Kalle

Deutsches Krebsforschungszentrum (DKFZ)
Nationales Centrum für Tumorerkrankungen (NCT) Heidelberg
Im Neuenheimer Feld 350
69120 Heidelberg

10 Mitglieder der Senatskommission für Grundsatzfragen der Genforschung

Prof. Dr. Jörg Hinrich Hacker
– Vorsitzender –
Bayerische Julius-Maximilians-Universität Würzburg
Institut für Molekulare Infektionsbiologie
Röntgenring 11
97070 Würzburg

Prof. Dr. Claus R. Bartram
Ruprecht-Karls-Universität Heidelberg
Institut für Humangenetik
Im Neuenheimer Feld 366
69120 Heidelberg

Prof. Dr. Herwig Brunner
Universität Stuttgart
Institut für Grenzflächenverfahrenstechnik
Nobelstraße 12
70569 Stuttgart

Prof. Dr. Bärbel Friedrich
Humboldt-Universität zu Berlin
Arbeitsbereich Mikrobiologie
Institut für Biologie
Unter den Linden 6
10099 Berlin

10 Mitglieder der Senatskommission

Prof. Dr. Werner Goebel	Bayerische Julius-Maximilians-Universität Würzburg Theodor-Boveri-Institut für Biowissenschaften (Biozentrum) Am Hubland 97074 Würzburg
Prof. Dr. Klaus-Peter Koller	Sanofi-Aventis Deutschland GmbH S&MA, General Affairs Valorisation & Innovation, H 831 Industriepark Höchst 65926 Frankfurt
Prof. Dr. Hans-Georg Kräusslich	Ruprecht-Karls-Universität Heidelberg Abteilung Virologie Im Neuenheimer Feld 324 69120 Heidelberg
Prof. Dr. Nikolaus Pfanner	Albert-Ludwigs-Universität Freiburg im Breisgau Institut für Biochemie und Molekularbiologie Hermann-Herder-Straße 7 79104 Freiburg
Prof. Dr. Renate Renkawitz-Pohl	Philipps-Universität Marburg Fachgebiet Entwicklungsbiologie und Parasitologie Karl-von-Frisch-Straße 8 35043 Marburg
Prof. Dr. Bettina Schöne-Seifert	Westfälische Wilhelms-Universität Münster Institut für Ethik, Geschichte und Theorie der Medizin Von-Esmarch-Straße 62 48149 Münster

10 Mitglieder der Senatskommission

Prof. Dr. Uwe Sonnewald Friedrich-Alexander-Universität
Erlangen-Nürnberg
Institut für Mikrobiologie,
Biochemie und Genetik
Lehrstuhl für Biochemie
Staudtstraße 5
91058 Erlangen

Prof. Dr. Jochen Taupitz Universität Mannheim
Institut für Deutsches,
Europäisches und
Internationales Medizinrecht
Schloss
68131 Mannheim

Prof. Dr. Gerd Utermann Institut für Medizinische Biologie
und Humangenetik
Schöpfstraße 41
6020 Innsbruck
Österreich

Prof. Dr. Riccardo Wittek Institut de Biotechnologie
Batiment de Biologie
Université de Lausanne
1015 Lausanne
Schweiz

Zuständiger Programmdirektor der DFG:

Dr. Frank Wissing Deutsche Forschungsgemeinschaft (DFG)
Kennedyallee 40
53175 Bonn

Development of Gene Therapy

Memorandum by the
Senate Commission on Genetic Research

Preface

After more than ten years, the Senate Commission on Genetic Research is issuing a second statement on gene therapy. Like hardly any other new form of treatment, gene therapy since the 1980s has repeatedly led to controversial discussions concerning its therapeutic potential and the associated health risks and ethical problems. In the meantime, however, comprehensive experimental research and initial clinical trials have significantly substantiated the until then only suspected therapeutic possibilities and their side effects. As the present statement shows, the realisation of the initially promised cures has made far less progress than expected. On the other hand, many of the sometimes exaggerated risks have also not been confirmed. Therapeutic successes as in congenital immunodeficiency diseases show that somatic gene therapy can be a valid treatment option. However, the observed and sometimes fatal side effects also show the present risks of this therapy. With regard to these problems, however, somatic gene therapy is not fundamentally different from other treatment options. The following holds: Before and during clinical testing and application, the benefits and risks must always be assessed in comparison with the alternative treatment options available and on the basis of solid information from experimental and clinical observations. In addition to being scientifically up-to-date, this requires close contact and an open exchange with scientists conducting basic research and those engaged in clinical research, as well as with the advisory ethics committees and the licensing authorities. All scientists are required to practice such an open discussion, just as the public is required to participate in this discus-

Preface

sion in a critical but unbiased manner. Therefore, this statement is addressed both to science and to the interested public and aims to point out the current state and the further perspectives of somatic gene therapy.

Since somatic gene therapy using retroviral vectors, despite some significant advances, still remains in a rather experimental stage, the Senate Commission rightfully warns against a broad application at the present time. In addition to further optimisation of the treatment, there is a significant need for research concerning our understanding of the observed and expected side effects. Therefore, it is welcome that the Deutsche Forschungsgemeinschaft (DFG, German Research Foundation) with its current funding of a special Priority Programme to examine the cell entry and persistence of gene therapy vectors has also made a significant contribution to understanding the mechanisms that can lead to possible side effects in somatic gene therapy. As was also confirmed by foreign experts, Germany has a number of internationally leading teams and very promising young researchers in this field. Therefore, it is even more important that these scientists, in addition to financial support and the provision of suitable clinical and scientific structures, are also given clearly visible career perspectives.

At this point, I would like to thank the Senate Commission on Genetic Research and the authors of this statement for giving a comprehensive presentation of the development of somatic gene therapy and for contributing to the ongoing discussion, on the basis of the current state of scientific knowledge and the future perspectives.

Bonn, December 2006

Professor Dr. Ernst-Ludwig Winnacker
President of the Deutsche Forschungsgemeinschaft

1 Summary

Since the Deutsche Forschungsgemeinschaft (DFG, German Research Foundation) first published a statement on gene therapy in 1995, this field has made enormous progress. Initial clinical successes have shown the therapeutic potential, but also the risks of this treatment option. Like all other forms of therapy, the clinical application of gene therapy requires a careful weighing of the benefits and risks. This field of research is still in a pilot stage and, like few other fields of medical research, relies on a close collaboration and a constant exchange between scientists engaged in basic research and clinicians of different disciplines. This requires an iterative optimisation process whereby, in contrast to classical clinical research, new results from the laboratory are directly transported to a clinical study and the knowledge gained from this process is then once again used to make further progress in the laboratory.

The successful further development of gene therapy over the medium-term especially requires research in the area of developing more efficient and safer vectors as well as the molecular investigation of the effects and side effects of gene therapy in the animal model and in the context of experimental clinical studies. A stronger focus has to be put on the assessment of the risk profile of potentially used therapeutic genes. The use of gene therapy is restricted to treatment and prevention. Gene doping or cosmetic application is not a justifiable use. The present legal framework is adequate; the current registration of gene transfer studies in a central registry should be continued. Overall, the past ten years have demonstrated the therapeutic potential of gene therapy

1 Summary

with initial successes in immunodeficiency diseases and have also delivered promising treatment approaches for a series of monogenetic hereditary diseases as well as for the acquired diseases of cancer and HIV infection/AIDS.

2 Introduction and brief historic outline

Gene therapy is defined as the introduction of genes into tissues or cells with the objective of obtaining a therapeutic or preventive benefit as a result of the expression and function of these genes. The process of introducing genes into cells is called **gene transfer**. A vehicle that carries the gene, the **vector**, is required to accomplish this. Therefore, gene therapy is the medical treatment with gene transfer drugs that are defined in the Medicines Act. The desired gene transfer exclusively concerns somatic cells (somatic gene therapy). Germline gene transfer is forbidden by law in Germany.

During the last two decades, only few fields of medical research have received as much attention as somatic gene therapy. Discoveries in molecular biology and genetics which in 2001 reached their peak with the sequencing of large parts of the human genome provided the prerequisites for gene therapy. In the early phase of gene therapy at the beginning of the 1990s these facts were partially discussed in an uncritical and too euphoric manner. This led to unrealistic expectations. The methods of gene transfer have significantly improved during the past few years, and the first clinically successful gene therapies have been performed in patients. Nevertheless, it is also necessary to emphasise today that the development of mature gene therapy procedures for many diseases that are otherwise untreatable will take many years, despite the fact that the success of individual gene therapies is foreseeable over the medium-term.

The first well documented gene therapy studies were started at the beginning of the 1990s. By 2005 it was estimated that more

2 Introduction and brief historic outline

than 1100 gene therapy studies have been conducted throughout the world; one third of these in Europe with a special focus in Germany. In 2003 as well as in November 2005 China approved the first gene therapy drugs for the treatment of certain malignant tumours. A first European application for the approval of a gene therapy drug for the treatment of an aggressive brain tumour was submitted to the European Agency for the Evaluation of Medicinal Products (EMEA) in 2005.

Despite continued great difficulties in the technical implementation, the successes of gene therapy can doubtlessly be confirmed today. For example, successful causal therapies have been developed during the past five years for patients with severe hereditary immunodeficiency diseases. These treatments are visibly beneficial to these patients with life-threatening conditions (see section 3). However, in one study, three children who were treated with retroviral vectors developed a leukaemia-like disease three years after treatment. At first, this was exclusively attributed to a side effect of the vectors used in combination with the underlying disease, but may possibly also be due to the overexpression of the target gene. This and the death of a patient in the USA in 1999 as a result of a very high systemically administered dose of adenoviral vectors were tragic events that were viewed by the public as a setback for gene therapy. Nevertheless, the same principles apply to gene therapy as to other medical interventions: Effective procedures are associated with potential side effects. Side effects can be reduced by improving the procedures when the underlying mechanisms are understood. For each indication and each procedure the relationship of effect to side effects (therapeutic index) must be determined in meticulous preclinical and clinical studies. German scientists have made important contributions in this field, from basic research of the vector-host interaction to clinical studies. Among other things, in 2006 they reported on the correction of a severe immunodeficiency in adult patients through gene therapy.

3 Clinical application: successes and setbacks

Compared with previous approaches, gene therapy offers a new way of treatment with a high potential for innovation since genes are used as drugs, while conventional drug development uses chemical substances, products from microorganisms or proteins. However, in practice it quickly became apparent that the initial prognoses for gene therapy had clearly underestimated the developmental requirements.

The selection of suitable vectors is crucial for the efficacy of a gene therapy. Different therapeutic or prevention objectives can require different vectors. For example, the selection depends on whether the gene transfer occurs in the patient (*in vivo*) or in the cell culture dish (*ex vivo*, or *in vitro*) since these procedures place different demands on the safety and targeting precision of the vector. The development of improved vectors for gene therapy continues to be one of the central tasks for research.

Gene transfer vectors for somatic gene therapy must have the following properties:

- They must be able to efficiently modify certain human cells.
- They must be able to ensure a sufficiently strong and sustained gene expression.
- They must exhibit a risk profile that is as low as possible for the desired form of treatment.

Presently, clinical gene therapy studies often use viral vectors that are incapable of replication and that are derived from retroviruses, adenoviruses, adeno-associated viruses and pox viruses. To this

3 Clinical application: successes and setbacks

purpose sections of the viral genome required for replication have been removed or inactivated and replaced by the therapeutic gene. After introducing the vectors into helper cell lines which contribute the necessary functions for virus formation, defective viruses are formed that are suitable for gene transfers, but which no longer can reproduce outside of the helper cells. Furthermore, plasmid DNA in pure form or mixed with other reagents is used as a non-viral vector. Viruses with a limited ability to reproduce are used in clinical studies in combination with therapeutic or preventive genes in cancer therapy and vaccination.

The disease groups that so far have mainly been examined in clinical studies on gene therapy are cancers, monogenic hereditary diseases, infectious diseases (especially HIV/AIDS) and cardiovascular diseases. Of these, cancers represent the largest segment (60 percent). Most clinical gene therapy studies are in very early clinical phases; only a few have reached phase III or have achieved proof of clinical efficacy. It must be assumed that many of the previously completed or presently ongoing clinical gene therapy studies of phases I and II will not yet lead to a drug that can be routinely used, since these are pilot studies for the treatment of very rare diseases.

Proof of the clinical efficacy of a gene therapy could particularly be obtained in studies on the treatment of severe immune defects. These are studies on the inherited combined immune defects ("severe combined immunodeficiency", X-SCID; the Cavazzano-Calvi and Fischer group in Paris as well as the Thrasher group in London) on adenosine deaminase deficiency (ADA-SCID; the Auiti and Bordignon group in Milan) and on chronic granulomatosis (CGD; the Grez and Hölzer group in Frankfurt/Main). First indications of the clinical efficacy of gene therapy were also obtained in the use of the CD40 ligand in chronic lymphatic leukaemia (Kipps group in San Diego), in the transfer of the GM-CSF gene in malignant melanoma (Lattime group in Philadelphia) as well as in the treatment of haemophilia B (factor IX; the McKay and High group in Stanford/Philadelphia).

Approximately ten years of intensive research and development conducted by many groups was required to treat a sufficiently large number of cells outside of the human body with gene vectors in such a manner that after returning these genetically modified cells to patients a therapeutic success could be rea-

3 Clinical application: successes and setbacks

lised. As expected, these initial successes in the treatment of monogenetic hereditary diseases were achieved with retrovirally modified blood stem or precursor cells, especially in the immune deficiency syndromes X-SCID and ADA-SCID. In these cases, *ex vivo* vectors derived from the murine leukaemia virus that are incapable of reproduction and result in the mainly accidental integration of the therapeutic gene into a chromosome of the particular host cell were used. The diseases mentioned offer especially favourable prerequisites for gene therapy: a therapeutic effect is already achieved when a limited number of target cells are modified, the genetically modified cells have a growth advantage in the organism or such an advantage is realised by pretreating the patients, and the cells modified with the gene transfer vector are not rejected because of the immunodeficiency.

Approximately three years after the first successful treatment of ten patients with X-SCID in the study conducted at Necker Hospital in Paris, three of these patients developed acute T-cell leukemias as a side effect of gene therapy and one of these patients died of the disease. In an exemplary international collaboration in which German scientists participated, significant progress was made in clarifying the molecular causes of this side effect. It was determined that the retroviral gene vectors used activated cellular proto-oncogenes during their integration into the genome of the treated T-cells and, thereby, contributed to the development of these cancers. Furthermore, more recent investigations cast doubts on the harmlessness of the correction gene used, since lymphomas, which were not necessarily related to the vector used, occurred during long-term studies in the animal model. Corresponding side effects so far have not occurred in a similar study of X-SCID patients at Great Ormond Street Children's Hospital in London. Therefore, the overall success of these gene therapy studies is positive. However, the recognised essential disadvantage of the vectors used in the Paris study with respect to the long-term course presently cannot be estimated with certainty because of inadequate clinical experience. In the treatment, which was carried out in Frankfurt/Main and reported in 2006, of adult patients with CGD by means of retroviral vectors, the integration of the vector genome into cell cycle-activated genes was also frequently observed. Although in this case, this integration presumably contributed to the therapeutic success

3 Clinical application: successes and setbacks

(by propagation of the successfully treated cells in the organism), in the meantime one of the patients treated in this study has died as a result of an infection-related complication after extensive functional loss of the treated cells. The other patients continue to benefit from the therapeutic effect. Taken together these results clearly show that a significant amount of research must still be undertaken regarding the connections between therapeutic efficiency and side effects in gene therapy.

Researchers in the field of gene therapy have from the start sought out public discussion and even in very early clinical trials have provided complete insight into the treatment risks and side effects. During this process, the public has not adequately perceived that, despite the leukemias described above, gene therapy of fatal diseases (such as monogenic hereditary immunodeficiency diseases) does not exhibit a higher rate of side effects than comparable conventional forms of treatment. So far, three of the 28 patients with X-SCID who were treated developed leukaemia as a side effect of the vectors used and one of these patients died. This corresponds to a side effect rate of about ten percent with a mortality rate of four percent. In conventional therapy of these same diseases with bone marrow or blood stem cell transplantation from HLA-identical family donors, the mortality rate is ten percent or higher. In the group of children selected for gene therapy (without HLA-identical donors from the family), the mortality rate at about 30 percent is even higher. However, it must be noted that the numbers mentioned rely on a small database. Leukaemia was already known as a possible risk before using gene therapy. However, the probability of its occurrence was unclear and was estimated as being low. After a conscientious evaluation and weighing of the risk, all participants decided to undergo therapy, because without gene therapy a long-lasting correction of the cell functions was not possible and the risk of the underlying disease significantly exceeded that of the therapeutic intervention.

In light of new information about the causes of the leukaemias, new and safer vectors have been developed in the meantime that should significantly reduce the risk of activating cellular oncogenes. Furthermore, sensitive preclinical models and diagnostic methods for toxicity determination have been described. In this way the molecular mechanisms of severe side effects in the future often can be avoided and detected earlier. Furthermore,

3 Clinical application: successes and setbacks

the newest data show that the toxic effect of the newly introduced gene in the X-SCID case possibly was underestimated. Therefore, in addition to the vector also the therapeutic gene must be more significantly included in the assessment of the risk profile. Based on more recent research, it has become even clearer that the previously used vectors very rarely caused severe side effects, but relatively frequently affected the expression of cellular genes. These vectors also affect the growth and function of gene-modified cells in an otherwise healthy organism. This was also shown in the CGD study described above. The risk of severe side effects probably depends on many additional factors and on the underlying disease; this can be analysed in the experimental model, but ultimately is not clearly recognisable until clinical testing. Therefore, both vector development and clinical implementation continue to require significant additional research.

The named successes of gene therapy were made possible by a significantly improved efficiency of the gene transfer. For example, hematopoietic cells today can be genetically modified with a high efficiency (>50 percent). However, achieving effective therapeutic levels was associated with a higher probability of symptomatic side effects. This simultaneously means that – as with other drug therapies – a further dose increase of the previously used vectors is more likely to be associated with more extensive side effects. Dose determination and toxicity determination have always been an inseparable part of developing active pharmaceutical substances and gene therapy is no exception. Through detailed investigation of the molecular causes, promising approaches have been discovered which will significantly improve the risk-benefit profile of the next generations of gene vectors. Therefore, for a series of monogenetic hereditary diseases and the acquired diseases of cancer and HIV infection/AIDS, gene therapy offers more innovative, worthwhile treatment options than ever before. It is especially the objective of HIV gene therapy to introduce protective genes (which, for example, prevent virus entry) into blood stem cells of the patient, whereby the successfully treated cells then continuously produce HIV-resistant immune cells, thereby, preserve the function of the immune system.

In contrast to the method presented above for diseases that are fatal if left untreated, the use of gene therapy vectors to

3 Clinical application: successes and setbacks

express antigens as a vaccine requires the use of vectors and procedures with a very low risk of side effects. Also in this case, the risk-benefit analysis must refer to the presently used vaccines against infectious diseases whose risk of side effects lies in the per mill range or below. Therefore, this application uses non-viral vectors or viral vectors that are incapable of reproduction, which effect only a temporary and local cell modification, but no chromosomal integration. The risk of these vectors is estimated as being very low, but the gene transfer efficiency is also lower and the gene expression is only temporary. Before each clinical trial of gene transfer drugs, a responsible risk-benefit analysis in terms of the individual approach used must be undertaken (as is also common with other drugs).

Because of the risks of gene therapy procedures with inserting vector systems as well as the necessary costs associated with their development, the use of retrovirally modified cells in humans is in the foreseeable future only permissible for the treatment of serious diseases after carefully weighing the particular risk-benefit relationship. Vector systems which do not cause permanent changes – and thus essentially do not differ from other pharmaceutically active substances – can also be used in diseases that are not life-threatening after comprehensive safety tests have been performed on the particular vector and the transgenic product. However, even with vectors that have been proven to be safe, an application of gene therapy that is not medically indicated, to improve performance, for example in competitive sports (gene doping), cannot be justified for ethical and medical reasons.

4 Present situation and further research needs

Research to develop a successful gene therapy serves as an example for other areas of *translational research*, in which knowledge obtained from biomedical basic research should be directly transferred to clinical application. This medical field faces special challenges and difficulties and its success to a large degree depends on a functioning dialog between the basic sciences (for example, vector development and optimisation) and applied clinical research. Of special significance for the successful development of gene therapy research in Germany is the close collaboration of scientists engaged in basic research and physicians experienced in treating the target disease. As is apparent from the results of previous clinical studies discussed above, the further development of promising gene therapy approaches requires an iterative optimisation process which, unlike prior clinical research, transports new knowledge from the laboratory into the clinic, tests it there, substantiates or discards a hypothesis and is then followed by renewed preclinical optimisations. Naturally, this must occur according to strict ethical criteria.

In Germany it is especially the targeted funding of the Federal Ministry of Education and Research (BMBF) and project funding by the DFG in the past ten years that have established a substantial number of successful and interdisciplinary gene therapy groups. Most notably, a number of younger natural scientists and physicians could be convinced to return to Germany from the USA. Especially in the fields of vector development and genome insertion analysis of retroviral vectors, but also in the implementation of clinical gene therapy studies, German scientists

4 Present situation and further research needs

have achieved an internationally recognised standing. The pioneering time of rapid successes in gene therapy research is coming to an end. Systematic basic research is increasingly required to solve the acknowledged problems. With appropriate funding the German research tradition of systematic and detailed analysis should serve as a good foundation for maintaining and expanding the achieved position over the long term. Research is especially needed in the following areas: (i) improved efficiency and safety of gene transfer vectors, (ii) optimising the specificity of viral vectors used for defined target cells in *in vivo* gene therapy, (iii) investigation of the persistence of gene-modified cells in patients and (iv) research of the molecular causes of side effects. In addition to these direct research topics, the aspect of vector production and testing also plays an important role for the development of gene therapy. Without adequate support it could prove limiting for further progress.

The first clinical successes of gene therapy when treating X-SCID were based on special preconditions: The defective bone marrow cells removed from these patients receive a selective advantage as a result of the therapeutic gene transfer; after being transplanted back into the patients, they respond to the natural signals of the body and can reproduce. These "healed cells" regenerate the immune system and even when only a relatively small number of cells are treated outside of the organism (*ex vivo*) this selection advantage results in a therapeutic success. In an analogous manner the treatment of CGD with gene therapy apparently gave rise to a selection advantage for the *ex vivo* treated cells by insertion of the vector genome into genes important for cell growth. However, in most diseases a selection advantage is not expected for the treated cells. Furthermore – as stated above – increase of the vector dose is possibly associated with an increased complication rate. As a consequence of this, additional research is needed in the development of more efficient and simultaneously safer vectors as well as in the investigation of the risk profile of therapeutically used genes, first *in vitro* and in the animal model and – if successful – in clinical studies. By better understanding the factors that can contribute to the multiplication of the viruses used, it is possible to achieve improvements, whereby over the medium-term additional optimisation can be expected by connecting viral functional complexes from different

4 Present situation and further research needs

viruses as well as in combinations with non-viral systems. In this way, it may be possible to combine desired properties of different vector systems and exclude undesirable properties. Against the background of the developing field of synthetic biology, the synthetic production of a gene transfer vector appears to be an attainable goal.

In addition to the fact that only relatively few cells can be treated in *ex vivo* gene therapy, an additional disadvantage – with the exception of blood cells – is that most somatic cells cannot be removed simply for cell culture therapy. Therefore, the objective is administration of the gene transfer vector in the patient (*in vivo*). During this process, the concentration of the administered vectors is very rapidly diluted and they come in contact with many cell types that are not affected by the particular disease. To become therapeutically effective and cause as few side effects as possible in other cells the *in vivo* applied vectors must be much more specific and effective when penetrating their particular target cells and in their therapeutic effect in these cells than in *ex vivo* gene transfer. Therefore, increasing the specificity of the vectors for therapeutically significant target cells of the organism and the efficacy of their penetration and effect in these cells is an additional research topic of central importance and an essential requirement for a broader clinical application of gene therapy.

Research continues to be required in investigating the molecular causes of the side effects of gene therapy. It is precisely this topic that has become extremely pertinent when three cases of leukaemia occurred in 28 successfully treated patients with X-SCID. The risk of side effects that can be caused by the mainly undirected integration of the retroviral vector genome into the genome of the host cell was originally estimated as being very small. Integration by chance can disturb the function of genes that normally regulate cell growth; undirected growth and tumours may result. Despite significant progress, the causes of the unexpectedly frequent occurrence of this side effect in X-SCID patients have not yet been fully clarified. Explaining the causes of side effects will not only lead to an understanding of the underlying molecular mechanisms, but also to more effective and safer gene transfer protocols and vectors for clinical application. More recent studies show that a change in cell proliferation through vector insertion possibly occurs more frequently, but that this does not necessarily lead to tumours. Investigating the underlying causes and conditions is of decisive importance for the further development of this approach in gene therapy. In addition to the untargeted integration of the vector, the therapeutic gene itself, can play a role in tumour development. Initial studies have shown that these effects occur with some delay and therefore do not become manifest in short-term experiments. Therefore, long-term preclinical observations must be performed and animal models must be developed that show corresponding side effects as early as possible.

A crucial factor for the clinical application of gene therapy is producing the required quantity of the gene transfer drug in accordance with the regulatory guidelines ("Good Manufacturing Practice" in manufacturing the investigational drug (GMP), "Good Laboratory Practice" in pharmacological/toxicological testing (GLP), "Good Clinical Practice" in clinical studies (GCP)). The development of gene therapy presently is in the hands of academic research groups and small biotechnology companies. Without question, university groups usually are not in a position to achieve vector production in accordance with GMP guidelines. Other countries have gone different ways with regard to GMP production of gene therapy vectors or genetically modified cells. In the USA university institutions have become involved with biotechnology companies and the NIH. From 1997 to 2003 the French foundation "l'Association Française contre les Myopathies" established a "Gene Vector Production Network" to make it easier to use and modify gene therapy vectors for research purposes. One of the ideas was to develop a European network from this for GMP production. However, this was not realised. In Great Britain the Department of Health has made available four million GBP for the time period 2003 to 2008 to be used in the production of gene therapy vectors for gene therapy studies within the NHS (National Health Systems). In Germany possibilities of GMP-consistent production of vectors for gene therapy are presently available at biotechnology companies, several pharmaceutical companies and at the Helmholtz Centre for Infection Research in Braunschweig. Fundamentally, vector production and testing can be realised either by establishing common production facilities in the public sector or through specialised small to medium-size businesses

4 Present situation and further research needs

(frequently as spin-offs of university groups) as service providers. In each case, adequate financing is required. Presently, these costs can only rarely be covered by project funding of the interdisciplinary academic research groups composed of physicians and natural scientists. In consideration of the fact that gene therapy can only continue to develop successfully in interaction between research and development, on the one hand, and in clinical studies, on the other hand, the aspect of vector production and testing is an important location factor for this field of research.

In summary, fundamental progress has been achieved in the successful application of gene therapy only as a result of intensive, basic science-oriented, preclinical and clinical research and testing as well as in constant communication between these disciplines. This need for research is acknowledged by the DFG, among other means, through the establishment of the Priority Programme "Mechanisms of Cell Entry and the Persistence of Gene Vectors" in 2005. The objective of this Priority Programme is the interdisciplinary investigation of the biological safety of entry and the persistence of viral and non-viral gene transfer vectors with a scientific focus on the cells of the hematopoietic and lymphatic system. This DFG Priority Programme which is more basic research-oriented complements other national ("innovative therapies" of the BMBF) and international (CLINIGEN of the EU) funding initiatives that transition to clinical application. Since these funding initiatives in each case only partially cover the necessary spectrum of basic and clinical research that is needed for translational research, close cooperation between the different funding organisations is essential to offer the interdisciplinary groups of scientists and clinicians funding that is of sufficient breadth. The programme for funding of clinical studies jointly conducted by the BMBF and DFG stands as an example for successful cooperation in this field. An alternative path is opened by the programme of the Clinical Research Units of the DFG in which, based on a clearly defined thematic focus, clinically relevant research fields are established in the clinics on a priority basis, also with the collaboration of scientists engaged in basic research. The close interaction between basic research scientists and clinicians within such an association guarantees optimal conditions for the requirements of translational research. An example of this is the Clinical

4 Present situation and further research needs

Research Unit "Stem Cell Therapy" at the medical school in Hannover, in which gene therapy approaches are also transferred to the clinic.

5 Legal and ethical considerations

Because of the associated unforeseeable risk potential, germline gene transfer in Germany is forbidden by the Embryo Protection Act for good reasons. Therefore, this will not be discussed in detail.

In comparison with other innovative therapies, somatic gene therapy does not pose any fundamentally different or new ethical or legal problems. Before a first application in humans, the associated risks must be clarified by means of an animal experiment. Furthermore, the Commission for Somatic Gene Therapy of the Scientific Advisory Board of the German Medical Council has also advised against individual treatments with gene transfer drugs, because the development of a therapy rationally only seems possible as a result of information obtained from its use in a series of subjects or patients during a clinical trial.

The production of organisms altered by genetic engineering in the laboratory as well as the establishment and operation of genetic engineering facilities are subject to compulsory registration or official approval in accordance with §§ 8ff. of the Genetic Engineering Act (GenTG). In contrast, the Genetic Engineering Act does not include the use of genetically altered organisms in humans. Therefore, the Medicines Act (AMG) should be applied primarily to a clinical study and the EU guideline 726/2004 should be applied to a subsequent approval. However, the treatment room used in a clinical study can be considered a genetic engineering facility in accordance with the Genetic Engineering Act.

As in the case of any other drug therapy that is still in the experimental stage, clinical studies that use gene transfer drugs

5 Legal and ethical considerations

require a voluntary and self-determined consent of the study participants. This consent must be preceded by adequate information provided by the physician which especially also must include indications of the measure's novelty and the expected or feared risks. Before and during the implementation of the clinical study, a risk-benefit analysis is required in which the risks of using the drug and the need to protect the target group (patients or healthy subjects), on the one hand, must be weighed against the possible benefit to the target group and the importance of the drug for medicine, on the other hand. This consideration, for example, leads to the use of gene transfer drugs with a low risk as prophylactic vaccines against infectious diseases in healthy subjects, while other gene transfer drugs for the treatment of fatal disease such as certain brain tumours are only tested on patients who have exploited all available means of conventional therapy and have a life expectancy of not more than a few months.

The coming into force of the 12th amendment of the Medicines Act (AMG) in 2004, by means of which the European GCP directive (Directive 2001/20/EU) was implemented, specifically establishes which guidelines and regulations should be applied to the manufacturing and development through market approval of gene therapy or gene transfer drugs (both terms are essentially synonymous) and defines the certain state of scientific knowledge. According to § 4 Par. 9 of the Medicines Act, this drug group on the one hand includes viral and non-viral gene transfer vectors, plasmid DNA and oncolytic viruses for *in vivo* gene transfer and, on the other hand, *ex vivo* genetically modified cells.

On the one hand, approval must be obtained in gene therapy for industrial or commercially produced individual formulations. For example, this affects gene transfer drugs that are produced by companies or also blood banks and which contain genetically modified cells according to a standard model while using the same gene vector and the same therapeutic gene. These drugs are intended for distribution to physicians for use in prevention, treatment or *in vivo* diagnostics in a specific patient. On the other hand, approval must be obtained for gene transfer drugs produced in advance for *in vivo* administration in many patients, such as, for example, viral vectors. Approval can be applied for on the basis of results from clinical studies of phases I to III at the European Agency for the Evaluation of Medicinal Products (EMEA).

5 Legal and ethical considerations

Before starting a clinical study, the positive evaluation of the competent ethics committee and the authorisation of the Paul Ehrlich Institute (PEI) are necessary. Insofar as none of the members have any relevant expertise, ethics committees will consult external experts in the evaluation of applications for gene therapy studies. The consultation that ethics committees in 1994 to 2005 usually obtained from the Commission on Somatic Gene Therapy of the Scientific Advisory Board of the German Medical Council has been discontinued until further notice by the Executive Board of the German Medical Association.

The 12th amendment of the AMG legally stipulates the approval or evaluation time limits of the Paul Ehrlich Institute and the competent ethics committees which expedites the procedures. In clinical studies of gene transfer drugs that contain genetically altered organisms (for example, viral vectors and viruses that to a limited degree are capable of reproduction or microorganisms), approval of the clinical study by the Paul Ehrlich Institute also includes the required approval to release. The same holds for the approval of gene therapy drugs by the EMEA.

Since gene transfer drugs are a new class of drugs that are subject to an ongoing development process, regulatory guidelines usually only can provide general information. Studies to confirm the quality, safety and efficacy of a given gene transfer drug usually must be decided on an individual basis. During this process, small biotechnology companies or academic research facilities are often confronted with new questions. In such a case, the applicant can make use of a consultation at the Paul Ehrlich Institute before submitting an application for a clinical study. The secrecy of the data and the confidentiality of the contents of the consultations are guaranteed by law.

Since 2004 a European database (EudraCT) has been established at EMEA which provides the competent authorities, the European Commission and the EMEA with necessary information about clinical studies in all European member states. However, this registry is not accessible to the public. In Germany there is an additional "German Registry for Somatic Gene Transfer Studies" (DeReG). This registry was established in 2001 on the initiative of the German Society of Gene Therapy (DG-GT) and the Commission for Somatic Gene Therapy in Freiburg and was funded by the BMBF. It includes information that presently cannot

5 Legal and ethical considerations

be obtained from any other available international study registry. For example, the Freiburg registry contains the side effects of individual patients from small phase I studies. Furthermore, this registry can be used to quickly and reliably inform the public, as needed (occurring side effects or successes). Therefore, the maintenance of such a publicly accessible registry appears to make sense to increase clarity and comprehensibility in the field of gene therapy. Therefore, the DFG continues to require that applicants show proof of having registered the gene therapy study in the DeReG before a clinical research project is approved. Whether the planned national study registry can fulfil the specified tasks in an analogous manner is presently still open and will not be known until its further implementation.

While the specific legal framework for gene therapy can be viewed as being adequate, the fundamental regulatory and structural problems for gene therapy research in a clinical environment are identical to those generally present in German academic clinical research. Basic conditions for clinical research in Germany that are more favourable overall, therefore, would also significantly improve the situation of scientists researching the clinical implementation of gene therapy. These problems and the corresponding solution proposals are described in the DFG White Paper on Clinical Research from 1999 as well as in the "Ten Key Points for Clinical Research" from 2004, which to a large extent are still relevant.

6 Conclusions and recommendations

- Since the first memorandum of the DFG on gene therapy in 1995, somatic gene therapy has achieved notable therapeutic successes, especially in monogenetically caused immunodeficiency diseases. In other potential applications it is still in early stages of development.

- Like every medical treatment, gene therapy also is associated with risks that must be carefully monitored and clarified with regard to their molecular causes. Clinical application requires a careful estimation of the benefits and risks for the specific indications. During this process, the public discussion of the successes and setbacks must also consider the prognosis of the underlying disease and alternative treatment options.

- The use of gene therapy is restricted to treatment and prevention. Use in gene doping or for cosmetic purposes is rejected.

- As long as safe retroviral vectors are unavailable, treatment with retroviral vectors should be limited to diseases without alternative treatment options.

- Furthermore, gene therapy requires a great deal of research. Basic research should be conducted in a direct, interdisciplinary manner that combines experiments in the animal model with clinical studies. Funding programmes such as the Clinical Research Units offer a basis for doing so and should increasingly be supplemented by funds that support translational research by the faculties. Furthermore, the necessary financing of expensive vector production in suitable facilities that have

6   Conclusions and recommendations

been approved in accordance with GMP guidelines must be ensured or be available through commercial suppliers.
- Current research should focus on the development of efficient and safe vectors for *in vitro* and *in vivo* application, including increased specificity for defined target cells as well as the molecular investigation of the effects and side effects. The risk profile of clinically used therapeutic genes must be considered.
- The existing basic legal conditions for gene therapy are sufficient.
- The registration of phase I and phase II gene transfer studies in a central registry has proven useful and continues to make sense. Registration should continue to be a requirement for funding by the DFG.

7 References

Baum C, Dullmann J, Li Z, Fehse B, Meyer J, Williams DA, von Kalle C: Side effects of retroviral gene transfer into hematopoietic stem cells. Blood. 2003 Mar 15; 101 (6): 2099–2114. Epub 2003 Jan 2. Review.

Guidance for Industry Gene Therapy Clinical Trials – Observing Participants for Delayed Adverse Events. FDA CBER.
www.fda.gov/Cber/gene.htm.

Hallek M, Buening H, Ried MU, Hacker U, Kurzeder C, Wendtner C-M: Grundlagen der Gentherapie [Foundations of Gene Therapy]. Internist 2001; 42: 1306–1313.

Kay MA, Glorioso JC, Naldini L: Viral vectors for gene therapy: the art of turning infectious agents into vehicles of therapeutics. Nat Med. 2001 Jan; 7 (1): 33–40. Review.

von Kalle C, Baum C, Williams DA: Lenti in red: progress in gene therapy for human hemoglobinopathies. J Clin Invest. 2004 Oct; 114 (7): 889–891.

Nabel GJ: Genetic, cellular and immune approaches to disease therapy: past and future. Nat Med. 2004 Feb; 10 (2): 135–141. Review.

O'Connor TP, Crystal RG: Genetic medicines: treatment strategies for hereditary disorders. Nat Rev Genet. 2006 Apr; 7 (4): 261–176. Review.

Nienhuis AW, Dunbar CE, Sorrentino BP: Genotoxicity of retroviral integration in hematopoietic cells. Mol There. 2006 Jun; 13 (6): 1031–1049. Epub 2006 Apr 19.

DFG-Denkschrift [White Paper] "Klinische Forschung" ["Clinical Research"]. 1999.
www.dfg.de/aktuelles_presse/reden_stellungnahmen/archiv/denkschrift_klinische_forschung.html

7 References

Internet addresses

German Registry for Somatic Gene Transfer Studies (DeReG)
   www.dereg.de
German Society of Gene Therapy e.V. (DG-GT)
   www99.mh-hannover.de/kliniken/zellth/dggt
European Society for Gene Therapy (ESGT)
   www.esgt.org
Paul Ehrlich Institute
   www.pei.de

8 Glossary

AAV vectors: Adeno-associated viruses that are used in gene therapy. They usually are not associated with human diseases, form stable particles and also infect resting cells, in which stable integration into the genome can occur. However, AAV particles only have a very limited capacity for taking up foreign genes. To propagate the AAV requires a second virus (a so-called helper virus, usually an adenovirus or parvovirus).

ADA-SCID: A hereditary, severe combined immune disease (SCID, severe combined immunodeficiency) in which the enzyme adenosine deaminase (ADA) is missing due to a gene defect. As a result, the body cannot degrade a protein that is poisonous to white blood cells and the T lymphocytes, which are important for the immune response, do not mature in the bone marrow or do so only in small numbers. Children affected by this disease are without any protection whatsoever and are almost completely at the mercy of all pathogens. Despite treatment and a life only under sterile conditions, they rarely survive their childhood.

Adenosine deaminase deficiency: See ADA-SCID.

Adenoviral vectors: Adenoviruses can be responsible, among other things, for the common cold in humans. Vectors with replication defects have a relatively high capacity for taking up genetic material. However, in higher doses they can lead to strong immune responses after administration.

8 Glossary

AMG: Act governing the production and distribution of medicines (Medicines Act). A current version can be found at www.gesetze-im-internet.de/amg_1976/index.html.

BMBF: Federal Ministry of Education and Research.

CGD: Chronic granulomatous disease, see chronic granulomatosis.

Chromosomal integration: Permanent integration of viral or introduced foreign genes into the chromosomes of the recipient.

Chronic granulomatosis: Genetically caused disorder of oxygen radical formation by phagocytes. As a result of the disturbed phagocyte function, the patients are strongly susceptible to infections and suffer from inflammatory diseases.

Clinical study of phases I, II, III and IV: Studies on the efficacy and toxicity of drugs in humans. These studies are subject to strict regulations. In phase I the toxicity or tolerance of new active substances is tested on a small number of healthy subjects. Based on the results of phase I, in phase II a larger number of study participants are used to determine the optimal dose. In phase III the actual effect is determined by using a sufficiently large number of patients with certain inclusion and exclusion criteria in order to obtain a statistically valid analysis. As required, this includes a comparison with a dummy medication without any active ingredients (placebo). Only based on a successful phase III study is the approval of a new drug possible. Thereafter, it is possible to further investigate or observe the effects of a new therapy in its approved form. This is a so-called phase IV study.

DeReG: German Registry for Somatic Gene Transfer Studies (www.dereg.de). This registry is publicly accessible.

DG-GT: German Society of Gene Therapy e.V.

EMEA: European Agency for the Evaluation of Medicinal Products (European Medicines Agency), drug approval authority of the EU.

EudraCT: EU-wide registry for clinical studies of the EMEA. This registry is not accessible to the public.

Ex vivo gene transfer: Gene transfer procedure in which the target cells (usually of the hematopoietic system) are initially isolated

8 Glossary

outside of the body and are then genetically altered with the vector and, if necessary, concentrated. Then these cells are once again administered to the body.

GCP: See Good Clinical Practice.

Gene expression: Implementation of the genetic information, usually in the form of proteins, to form cell structures and signals.

Gene therapy: Treatment involving the introduction of genes into the tissues or cells with the objective of obtaining a therapeutic or preventive benefit as a result of the expression and function of these genes.

Gene transfer: The methodical procedure of introducing genes into cells.

GenTG: Genetic Engineering Act. A current version can be found at www.gesetze-im-internet.de/gentg/index.html.

Germline gene transfer: Gene transfer in germ cells (egg or sperm cells or their precursors). Alterations of the genotype would also be passed on to successor generations. Germline transfer is legally prohibited in Germany.

GLP: See Good Laboratory Practice.

GM-CSF: Granulocyte-macrophage colony-stimulating factor. A so-called cytokine which stimulates the growth of macrophages and, thereby, can induce an immune response against certain types of skin cancer.

GMP: See Good Manufacturing Practice.

Good Clinical Practice: International rules for preparing and conducting clinical studies in accordance with ethical and practical considerations based on the present state of scientific knowledge. Additional details can be found at www.emea.eu.int/pdfs/human/ich/013595en.pdf.

Good Laboratory Practice: International rules and standards for quality assurance of the organisational processes and conditions of non-clinical health and environmental tests. Additional details can be found at http://ec.europa.eu/enterprise/chemicals/legislation/glp/index_en.htm.

8 Glossary

Good Manufacturing Practice: International rules and standards for quality assurance in the manufacturing of medical devices and active substances. Additional details can be found at www.emea.eu.int/Inspections/GMPhome.html.

In vivo gene transfer: In contrast to *ex vivo* gene transfer (see above), the gene vectors in this case are directly introduced into the body of the patient. Depending on the cell specificity of the vector used, the infection or the integration of the foreign gene then occurs in a more or less targeted manner in certain cell types.

Monogenic hereditary diseases: Diseases that are caused by the alteration of a single gene.

Oncogene: A gene that usually plays a role in cell cycle regulation and whose activation through mutation contributes to or causes the development of cancer.

Oncolytic viruses: Viruses that can infect and switch off tumour cells in a targeted manner.

Plasmid DNA: DNA that is not integrated in a genome, but instead is present as an independent, ring-shaped structure in a cell. This is usually not duplicated in cell division and thus dissipates after several cell divisions – unless the plasmid DNA is permanently integrated into the genome.

Proto-oncogene: A gene which through a mutation can be altered to an oncogene (see above).

Retroviral vectors: Gene vectors that originate from retroviruses. Retroviruses are RNA viruses. However, their RNA genome is transcribed to DNA and permanently integrated into the genome of a cell. Retroviral vectors based on murine leukaemia viruses infect many different cell types, partially with a very high efficiency. But they cannot infect cells that are not actively dividing (for example, nerve cells). However, this can be achieved by using HIV-based lentiviral vectors.

Somatic cells: Body cells whose genetic information cannot be inherited by successor generations. They form the majority of human cells; only germ cells (egg and sperm cells) can transfer genetic information to the next generation and form the so-called germline (see above).

8 Glossary

Somatic gene therapy: Application of gene transfer to somatic cells (see below). Genetic alterations are not passed on to offspring.

T-cell leukaemia: Blood cancer in which the regulation of white blood cell (T cells) reproduction is out of control and leads to a flooding of the blood and lymphatic system with degenerated cells.

Therapeutic index: The therapeutic index (also therapeutic window or therapeutic quotient) of a drug describes the ratio of its therapeutic to its toxic dose. The larger the therapeutic index, the less dangerous is the drug.

Vector: A vehicle that transports a therapeutic gene into the cells of the patient. In addition to different viruses that are mostly incapable of reproduction, plasmid DNA (see above) is used, as a non-viral vector, either in pure form or mixed with other reagents.

X-SCID: A hereditary, severe, combined immune disease (SCID, severe combined immunodeficiency). Due to the mutation of a gene for a common building block of several different types of interleukin receptors, no defence cells of the immune system can be formed. Affected patients – usually children – are highly susceptible to infections. The underlying gene is located on the X chromosome which explains the designation X-SCID.

9 Members of the Working Group on "Gene Therapy", who prepared the present statement

As members of the Senate Commission on Genetic Research:

Prof. Dr. Hans-Georg Kräusslich
– Chair –

Ruprecht-Karls-Universität Heidelberg
Abteilung Virologie
Im Neuenheimer Feld 324
69120 Heidelberg

Prof. Dr. Claus R. Bartram

Ruprecht-Karls-Universität Heidelberg
Institut für Humangenetik
Im Neuenheimer Feld 366
69120 Heidelberg

Prof. Dr. Jochen Taupitz

Universität Mannheim
Institut für Deutsches, Europäisches und Internationales Medizinrecht
Schloss
68131 Mannheim

9 Members of the Working Group on "Gene Therapy"

As external experts:

Prof. Dr. Klaus Cichutek
Paul-Ehrlich-Institut (PEI)
Paul-Ehrlich-Straße 51–59
63225 Langen

Prof. Dr. Charles Coutelle
Imperial College London
Faculty of Life Sciences
Division of Cell and Molecular Biology
Wolfson Biochemistry Building
Exhibition Road, South Kensington
London SW7 2AY
Großbritannien

Prof. Dr. Michael Hallek
Universität zu Köln
Klinik I für Innere Medizin
Hämatologie und Onkologie
Joseph-Stelzmann-Straße 9
50931 Köln

Prof. Dr. Christof von Kalle
Deutsches Krebsforschungszentrum (DKFZ)
Nationales Centrum für Tumorerkrankungen (NCT) Heidelberg
Im Neuenheimer Feld 350
69120 Heidelberg

10 Members of the Senate Commission on Genetic Research

Prof. Dr. Jörg Hinrich Hacker – Chair –	Bayerische Julius-Maximilians- Universität Würzburg Institut für Molekulare Infektionsbiologie Röntgenring 11 97070 Würzburg
Prof. Dr. Claus R. Bartram	Ruprecht-Karls-Universität Heidelberg Institut für Humangenetik Im Neuenheimer Feld 366 69120 Heidelberg
Prof. Dr. Herwig Brunner	Universität Stuttgart Institut für Grenzflächen- verfahrenstechnik Nobelstraße 12 70569 Stuttgart
Prof. Dr. Bärbel Friedrich	Humboldt-Universität zu Berlin Arbeitsbereich Mikrobiologie Institut für Biologie Unter den Linden 6 10099 Berlin

10 Members of the Senate Commission on Genetic Research

Prof. Dr. Werner Goebel
Bayerische Julius-Maximilians-Universität Würzburg
Theodor-Boveri-Institut für Biowissenschaften (Biozentrum)
Am Hubland
97074 Würzburg

Prof. Dr. Klaus-Peter Koller
Sanofi-Aventis Deutschland GmbH
S&MA, General Affairs
Valorisation & Innovation, H 831
Industriepark Höchst
65926 Frankfurt

Prof. Dr. Hans-Georg Kräusslich
Ruprecht-Karls-Universität Heidelberg
Abteilung Virologie
Im Neuenheimer Feld 324
69120 Heidelberg

Prof. Dr. Nikolaus Pfanner
Albert-Ludwigs-Universität Freiburg im Breisgau
Institut für Biochemie und Molekularbiologie
Hermann-Herder-Straße 7
79104 Freiburg

Prof. Dr. Renate Renkawitz-Pohl
Philipps-Universität Marburg
Fachgebiet Entwicklungsbiologie und Parasitologie
Karl-von-Frisch-Straße 8
35043 Marburg

Prof. Dr. Bettina Schöne-Seifert
Westfälische Wilhelms-Universität Münster
Institut für Ethik, Geschichte und Theorie der Medizin
Von-Esmarch-Straße 62
48149 Münster

10 Members of the Senate Commission on Genetic Research

Prof. Dr. Uwe Sonnewald
Friedrich-Alexander-Universität
Erlangen-Nürnberg
Institut für Mikrobiologie,
Biochemie und Genetik
Lehrstuhl für Biochemie
Staudtstraße 5
91058 Erlangen

Prof. Dr. Jochen Taupitz
Universität Mannheim
Institut für Deutsches,
Europäisches und
Internationales Medizinrecht
Schloss
68131 Mannheim

Prof. Dr. Gerd Utermann
Institut für Medizinische Biologie
und Humangenetik
Schöpfstraße 41
6020 Innsbruck
Österreich

Prof. Dr. Riccardo Wittek
Institut de Biotechnologie
Batiment de Biologie
Université de Lausanne
1015 Lausanne
Schweiz

Responsible Programme Director of the DFG:

Dr. Frank Wissing
Deutsche Forschungsgemeinschaft (DFG)
German Research Foundation
Kennedyallee 40
53175 Bonn